MW00837354

Chaos in the Heavens

Chaos in the Heavens

The Forgotten History of Climate Change

Jean-Baptiste Fressoz and Fabien Locher

Translated by Gregory Elliott

VERSO
London • New York

This book is supported by the Institut français (Royaume-Uni) as part of the Burgess programme.

This work was published with the help of the French
Ministry of Culture – Centre national du livre
Ouvrage publié avec le concours du Ministère français
chargé de la culture – Centre national du livre

La traduction de ce livre a bénéficié d'une subvention du Centre de Recherches Historiques de l'EHESS (UMR 8558) et d'une subvention du Labex TEPSIS (Transformation de l'Etat, politisation des sociétés et institution du social)
The translation of this book has benefitted from a grant from the Centre de Recherches Historiques de l'EHESS (UMR 8558) and a grant from the Labex TEPSIS (Tranformation de l'Etat, politisation des sociétés et institution du social)
First published by Verso 2024
Translation based on the revised French edition of *Les révoltes du ciel. Une histoire du changement climatique (xve–xxe siècle)*

1 3 5 7 9 10 8 6 4 2

Verso
UK: 6 Meard Street, London W1F 0EG
US: 388 Atlantic Avenue, Brooklyn, NY 11217
versobooks.com

Verso is the imprint of New Left Books

ISBN- 13: 978-1-83976-722-7
ISBN- 13: 978-1-83976-724-1 (US EBK)
ISBN- 13: 978-1-83976-723-4 (UK EBK)

British Library Cataloguing in Publication Data
A catalogue record for this book is available from the British Library

Library of Congress Cataloging-in-Publication Data

Names: Fressoz, Jean-Baptiste, author. | Locher, Fabien, author. | Elliott, Gregory, translator.
Title: Chaos in the heavens : a history of climate change from the fifteenth to the twentieth century / Jean-Baptiste Fressoz and Fabien Locher ; translated by Gregory Elliott.
Other titles: Révoltes du ciel. English
Description: London ; New York : Verso, 2024. | 'Translation based on the revised French edition of Les révoltes du ciel : une histoire du changement climatique (xve-xxe siècle).' | Includes bibliographical references and index.
Identifiers: LCCN 2023042118 (print) | LCCN 2023042119 (ebook) | ISBN 9781839767227 (hardback) | ISBN 9781839767241 (US EBK) | ISBN 9781839767234 (UK EBK)
Subjects: LCSH: Climatic changes – History. | Ecology – History. | Global warming – History.
Classification: LCC QC903 .F73713 2023 (print) | LCC QC903 (ebook) | DDC 551.609 – dc23/eng/20231102
LC record available at https://lccn.loc.gov/2023042118
LC ebook record available at https://lccn.loc.gov/2023042119

Typeset in Minion Pro by MJ&N Gavan, Truro, Cornwall
Printed and bound by CPI Group (UK) Ltd, Croydon CR0 4YY

To Cecilia, Leonor and Esteban

To Faustine, Arsène and Hadrien

Contents

Introduction

Ten Theses on Climate Change

London, Palace of Whitehall, February 1662. King Charles II of England has summoned his closest advisors. A man comes before him, bows, and swears allegiance. He carries an enormous weight on his shoulders: his whole world is at stake in this royal audience. John Winthrop is governor of the colony of Connecticut: the title is impressive, but who does he speak for? Several thousand men, women, and children. A people without rights, for, in the absence of a royal charter, they have no legal existence. Winthrop has come in search of such recognition.

For his part, the king has plans for America. He wants to get to know it better and administer the lands on the edge of his kingdom more efficiently. Robert Boyle backs him in this task. He is one of the great scientists of his time: the father of experimental physics. He wants to enlist the new sciences in the service of England and Empire. He is present at the scene in 1662.

Charles II listens to Winthrop and then questions him in turn: he enquires about the prevailing temperature in the colonies. As recounted by Boyle, Winthrop seizes the opportunity and announces that it has changed; that the cold has receded with colonization. Before his king, he makes three inseparable claims: that England is sovereign over these lands; that the colonists, who are virtuous and persevering, have set nature to work; and that divine Providence has hallowed the imperial enterprise. Here, climate change is a form of optimism.

Connecticut would have its charter. Winthrop became a member of the Royal Society – the holy of holies of modern science whose architect was Boyle. It witnessed debates over the New World, the nature of matter, the globe, the animal world, empire, profit and reason. But anxiety was already casting a shadow over hopes of conquest. From Barbados, a traveller warned the scientists back in London: if we cut down trees, colonization, rather than making the climate milder, could dry it to the point of desertification. Was England not similarly threatened? Holistic thinking made such concerns meaningful. In the great providential water cycle, trees interceded between clouds, rivers, and oceans: their loss might betoken a hopelessly dry sky. Conquering Earth but losing the sky: the transformation of nature acquired an aftertaste of anxiety.

This book's first thesis possesses a kind of self-evidence: European societies did not experience centuries of dramatic fluctuations in climate, or live through the Little Ice Age,[1] without caring about the development of the climate. On the contrary. From the dawn of the modern age to the start of the twentieth century, a host of scientists, statesmen, colonists, historians, agronomists, and engineers debated climate change endlessly, both rejoicing at it and being worried by it. While they rarely used the word 'climate',[2] they all referred to phenomena that would today

1 Historical climatology has studied this cold sequence, which ran from the beginning of the fourteenth century to the end of the nineteenth, at length, from the foundational works of Emmanuel Le Roy Ladurie and Hubert Lamb onwards. See Ladurie, *Histoire du climat depuis l'an mil*, Paris: Flammarion, 1967 and *Histoire humaine et comparée du climat*, 3 vols, Paris: Fayard, 2004–09; Lamb, *Climate: Present, Past and Future*, 2 vols, New York: Routledge, 2001(first published 1972 and 1977). Among the rich recent literature linking historical climatology with social, economic and political processes, see, in particular, Georgina H. Endfield, *Climate and Society in Colonial Mexico: A Study in Vulnerability*, Oxford and Malden: Blackwell, 2008; Emmanuel Garnier, *Les Dérangements du temps. 500 ans de chaud et de froid en Europe*, Paris: Plon, 2009; Geoffrey Parker, *Global Crisis: War, Climate Change and Catastrophe in the Seventeenth Century*, New Haven: Yale University Press, 2014; Sam White, *The Climate of Rebellion in the Early Modern Ottoman Empire*, Cambridge: Cambridge University Press, 2011; Dagomar Degroot, *The Frigid Golden Age: Climate Change, the Little Ice Age, and the Dutch Republic, 1560–1720*, Cambridge: Cambridge University Press, 2019.

2 From the sixteenth to the nineteenth century, the meaning of the term 'climate' was multiple and variable. It was initially used in its primary sense, derived from Greek cosmography, to refer to the portion of the terrestrial globe divided by two lines of latitude. It was

be characterized as climatic: rains that had dried up, winds that had changed, hitherto unknown cold or warm weather. They asked themselves: are these changes spontaneous or – as they often thought – the effects of human activity?

For this is our second thesis: belief in the climatic impact of *human action* has profoundly marked European societies over the long term. Our ancestors long thought they could change the rains, the temperature and the winds. But this had nothing to do with contemporary global warming. They were concerned with neither CO_2 nor the greenhouse effect. Instead they thought that clearing forests, transforming the planet's soil and plant cover, would alter the water cycle and hence the climate.

Countless historical documents speak to us of this power, its promises and its dangers. It is invoked whenever history takes great strides forward: by the first conquistadors, by nineteenth-century imperialists, by the French or American revolutionaries of the late eighteenth century.

We have sought to attend to these voices full of hope and anxiety. Not in order to seek in them the 'origin' of our present-day climate concerns,[3] but in order to write the history of a 'reflexivity' – a way of thinking about our relationship to the environment – that profoundly marked European societies for more than four centuries.[4]

---

also employed very generally as synonymous with region, country, land. Then, from the beginning of the eighteenth century, the notion of 'climate' as a space distinguished from other spaces by the prevalent cold or heat, dryness or humidity there, became established. It is this latter sense which, quantified by scientific meteorology, was to evolve into the contemporary idea of climate as the set of atmospheric features characteristic of a place or region. Added to this is the medical (neo-Hippocratic) notion of climate.

3 Spencer R. Weart, *The Discovery of Global Warming* [2003], Cambridge, MA and London: Harvard University Press, 2008.

4 This investigation is situated in a fast-growing field of research studying the role of climate in the life of past societies and prioritizing the viewpoint of the relevant actors. We may mention Richard Grove, *Green Imperialism: Colonial Expansion, Tropical Island Edens and the Origins of Environmentalism, 1600–1860*, Cambridge: Cambridge University Press, 1995 and *Ecology, Climate and Empire: Colonialism and Global Environmental History, 1400–1940*, Cambridge: White Horse Press, 1997; James R. Fleming, *Historical Perspectives on Climate Change*, New York and Oxford: Oxford University Press, 1999; Jan Golinksi, *British Weather and the Climate of Enlightenment*, Chicago: University of Chicago Press, 2007; Diana K. Davis, *Resurrecting the Granary of Rome: Environmental History and French Colonial Expansion in North Africa*, Athens: Ohio University Press, 2007; Anoushka Vasak, *Météorologies. Discours sur le ciel et le climat des Lumières au romantisme*, Paris: H. Champion, 2007; Vladimir Jankovic, *Confronting the Climate: British Airs and the Making of Environmental Medicine*, Basingstoke: Palgrave Macmillan, 2010; James R. Fleming and Vladimir Jankovic, eds, *Osiris*, vol. 26, no. 1, 2011, special

This profound imprint is commensurate with what climate represented at the time. Its importance was crucial for the agrarian societies of the past. Firstly, because the threat of a bad season constantly loomed over harvests and human beings' ability to feed themselves. The spectre of famine and riot made the climate an incendiary political issue, which destabilized – and sometimes overthrew – political regimes. Secondly, because the climate was not merely a matter of grain but was long regarded as the fundamental matrix of nature, as what moulded all living beings, humans included, determining their form, health and characters.[5] Acting on the climate, therefore, held out the dizzying prospect of creating new new types of human being.

Above all, this action took the form of altering forest cover. Forests obviously played a key role in past societies. Wood was long the primary

issue on 'Klima'; Fredrik Albritton Jonsson, *Enlightenment's Frontier: The Scottish Highlands and the Origins of Environmentalism*, New Haven: Yale University Press, 2013; Diana K. Davis, *The Arid Lands: History, Power, Knowledge*, Cambridge, MA: MIT Press, 2016; Caroline Ford, *Natural Interests: The Contest over Environment in Modern France*, Cambridge, MA: Harvard University Press, 2016; Gillen d'Arcy Wood, *L'Année sans été. Tambora, 1816. Le volcan qui a changé le cours de l'histoire*, Paris: La Découverte, 2016; Anya Zilberstein, *A Temperate Empire: Making Climate Change in Early America*, New York: Oxford University Press, 2016; Sara Miglietti and John Morgan, eds, *Governing the Environment in the Early Modern World: Theory and Practice*, London and New York: Routledge, 2017; Deborah R. Coen, *Climate in Motion: Science, Empire, and the Problem of Scale*, Chicago: University of Chicago Press, 2018; Victoria C. Slonosky, *Climate in the Age of Empire: Weather Observers in Colonial Canada*, Boston: American Meteorological Society, 2018. Meanwhile, we should not forget older but important works by Clarence J. Glacken, *Traces on the Rhodian Shore: Nature and Culture in Western Thought from Ancient Time to the End of the Eighteenth Century*, Berkeley: University of California Press, 1967 and Jean Ehrard, *L'idée de nature en France dans la première moitié du XVIIIe siècle*, Paris: SEVPEN, 1963, vol. 2.

5 In the neo-Hippocratic tradition long dominant in medicine, it was the set of 'things around us', in particular 'waters, airs and places', that determined the health or illness of animals and humans. 'Climate' was used to describe these influences. But what matters to us is that the meteorological characteristics of places were regarded as preponderant among them. On neo-Hippocratic medicine, see, in particular, the classic study by Jean-Paul Desaive, Jean-Pierre Goubert, Emmanuel Le Roy Ladurie, Jean Meyer, Otto Muller and Jean-Pierre Peter, *Médecins, climat et épidémies à la fin du XVIIIe siècle*, Paris: Mouton La Haye, 1972, and Andrea A. Rusnock, *Vital Accounts: Quantifying Health and Population in Eighteenth-Century England and France*, Cambridge: Cambridge University Press, 2002.

Recent works explore the links between milieus, races and domination: see Claude-Olivier Doron, *L'Homme altéré: races et dégénérescence (XVIIe–XIXe siècle)*, Paris: Champ Vallon, 2016 and Ferhat Taylan, *Mésopolitique. Connaître, théoriser et gouverner les milieux*, Paris: Sorbonne Éditions, 2018.

source of energy and an irreplaceable material for shipping, transport, building, tools, and many other things besides. Forests represented an immense capital, as well as being habitats and means of subsistence; they occupied a central place in imaginaries and sensibilities. Trees imbued climate change with an enormous political charge.

The book's third thesis is that anthropogenic climate change was, over the long term, an intellectual and practical framework in the service of European imperial expansion. With the 'discovery' of America, the idea took root that colonization was also a climatic normalization, a way of improving the continent's climate by clearing and cultivating land. It was a promise to the colonists and a discourse of domination: a way of saying that native peoples had never really owned the New World. In the eighteenth century, acting on the climate served to rank societies and their historical trajectories hierarchically: Amerindian peoples still in the infancy of a savage climate; European peoples creating the mild climate of their continent; Oriental peoples destroying theirs. The Maghreb, India and, later, Black Africa: in the nineteenth and twentieth centuries, the French and British empires were built on accusing Blacks and Arabs, Islam, nomadism, and the 'primitive' mentality, of wrecking the climate.[6] Colonization was conceived and presented as an attempt to restore Nature. The white man must mend the rains, make the seasons milder, push back the desert – and to that end command the natives. Discarded in the late nineteenth century in Europe, the thesis of anthropogenic climate change held sway for longer in those empires where, four centuries earlier, it had taken root.

Our fourth thesis is that science long ago seized on the issue of climate changes. From the time of Boyle and Winthrop, meteorological instruments began to be used to track potential anthropogenic climate change. The physiology of plants was invented in concert with the issue of disruptions to the water cycle. The study of climate change was systematized at the end of the eighteenth century: scientists analysed the old meteorological records, consulted historical sources, studied the evolution of rivers, vegetation and glaciers. In short, they laid the foundations of contemporary historical climatology. In the nineteenth century, these bodies of knowledge became full-fledged political weapons. The authorities deployed them to reassure people prey to the torments of cold and hunger: the bad season was only temporary and local shortages would disappear by themselves if

6 See especially Davis, *Resurrecting the Granary of Rome*.

the market was abided by. The political order could be guaranteed because nature was fundamentally stable. At the same time, the science of the climatic past played a key role in intense struggles over forests, property, and the regulation of nature. In France, climatology was invoked in Parliament and the press; the state mobilized its whole apparatus to understand the climatic trajectory of its territory. Meteorology, botany, agronomy, geology, medicine, physics, history, Oriental studies: this book unearths a whole hidden side of the history of knowledge, dealing with shifts in climate and human responsibility for the changes in it.

Changes, but on what scale? The contemporary age – fifth thesis – does not enjoy a monopoly on the global. The issue of climate change was broached on a planetary and continental scale since the inception of the modern age. In the eighteenth century, Buffon, one of the most famous scientists of his time, described an Earth growing colder and piling up ice at the poles. Scientists and travellers did not hesitate to link the signs of cooling observed on five continents to the destiny of the planet. In the theory of the water cycle, too, the Earth was already conceived as a unified whole, a system of material flows connecting soil, ocean, atmosphere, and vegetation. For its part, experimental science attempted via small-scale experiments to grasp the dynamics of the planet and climate. In the nineteenth century, theories of climatic deterioration described the mechanisms which, from the mountains to the poles, heralded global disasters. In this period, chemists and geologists identified the main self-regulating mechanisms conducive to the stability of the atmosphere's composition. Such visions were not exclusive to scientists: when questioned, French Restoration elites spoke of climatic influences on a French, European and world scale. The commonplace idea that global change and the contemporary sciences of the 'Earth system' constitute an ontological revolution, a radical transformation in our relationship to the planet, stems from neglect of this history.[7]

Our sixth thesis has a more historiographical tonality. Recent research remains marked by the problem of the origin of 'environmental awareness'.[8]

---

7 James Lovelock, *Gaia: A New Look at Life on Earth*, Oxford: Oxford University Press, 1979; Hans Joachim Schellnhuber, 'Earth System Analysis and the Second Copernican Revolution', *Nature*, vol. 402, 1999, C19–C23; Clive Hamilton and Jacques Grinevald, 'Was the Anthropcene Anticipated?', *The Anthropocene Review*, vol. 2, 2015, pp. 1–14.

8 See Ford, *Natural Interests* and Paul Warde, *The Invention of Sustainability: Nature and Destiny, c. 1500–1870*, Cambridge: Cambridge University Press, 2018.

In a highly influential book, the historian Richard Grove identified its point of emergence in the climatic anxieties of some eighteenth-century colonial administrators.[9] Upon extending the inquiry over four centuries, however, it turns out that the issue of 'awareness' is not the appropriate one: it generates teleological narratives, reaches ever further back into the past, and imprisons us in an intellectual genealogy. What we document is different: not an origin, but the rise, within old conceptions of the climate mixing optimism and pessimism, of an image of *collapse* that gained the upper hand *as a result of the French Revolution*. Neither discovery nor invention, then, but a process of *politicization* and *dramatization* in a specific space and time: the France of 1789 and the troubled political decades of the nineteenth century. This image of climate collapse would subsequently circulate on a world scale throughout Europe, America, and the empires.

As we can see, France occupies a very special place in the global and long-term history of climate change. This is our seventh thesis. In no other national context do we find so many debates, anxieties and reflections prompted by a possible deterioration in climates caused by man. Elsewhere, the effects of climate action were far from being so dreaded – when indeed they were not regarded as a blessing. This is not to invoke a cultural peculiarity or a discovery attributable to French science. France's uniqueness emerged from neither the literary sphere nor the scientist's study. It was the result of merciless struggles between factions of the Revolution, the Restoration, moderate monarchy, and economic liberalism. In this book, climatic deterioration does not – or, at least, does not only – appear in scholarly discourse or inspired prose, but through the voices of mayors, property-owners, deputies, engineers, pharmacists, doctors and journalists, administrators and foresters. It is not an intellectual game: it is a country crippled by anxiety.

Eighth thesis: the idea of climate collapse became a tool for controlling ordinary people's uses of nature. It formed the centrepiece of a discourse on the natural order that served to discipline these uses following the end of feudalism. The peasants of Year II stood accused: in not respecting private property, they had destroyed the forests and hence the climate. They had to learn not to misunderstand the true meaning of the French Revolution. It is no accident that the idea of climate collapse and Malthus's theory were contemporaneous. In both cases nature was enlisted into a project for

9 Grove, *Green Imperialism*.

governing the masses – their fertility in England, and their appetite for land and wood across the Channel. Thus, anthropogenic degradation was soon brandished by a forestry technocracy endowed with immense power in the nineteenth century. It was the strong arm of the state in the countryside, regulating access to a vital resource. In France, rural populations were singled out for their archaic practices, destructive of forests and hence of the climate. In Algeria and throughout the Maghreb, foresters were to the fore in accusing the '*Arabe*' of being '*l'ennemi de l'arbre*'. In the name of climate and forest conservation, they waged a crusade for 'modernizing' the uses of nature that targeted the ways of life and, sometimes, the very existence of communities.

Debates over the climatic threat are also – ninth thesis – intimately bound up with the expansion of liberal capitalism in nineteenth-century France. The problem of climate change, as we have seen, constantly interfered with forestry policy. By the same token it became bound up with basic political and economic questions. How was the enormous capital represented by forests to be managed? How should it be increased and put to profitable use to boost the public finances, to serve as security for loans, or to stimulate trade? And how to do all this without risking compromising the climate, and thus the country's prosperity and political stability? Zealots for the market and absolute property rights fought to free forest ownership from any restrictions. But if trees regulate the climate, should not the state regulate, prohibit, constrain; protect and conserve the vast forests under its control? These questions gave rise to intense struggles in parliamentary arenas, the media, and the corridors of ministries. At stake was the compatibility between market society and the integrity of ecological balances: could the management of nature be left to the goodwill of property-owners? Did not the fundamental arbitration between forest and cultivation, and hence between energy and grain, belong to the nation as a whole? The issue of *temporality* was also central: for the pro-regulation camp, only the state was capable of long-term thinking and hence of conserving. Supporters of an outright return to monarchy were similarly obsessed. But they believed it was the vocation of the king, the nobles and the clergy to act as guarantors of nature's longevity. At the very moment when modern institutions of political representation were being invented, they were the site of debates about natural cycles and the need to consider all the possible long-term repercussions of our actions on climate and the environment.

Our tenth and last thesis is that, from the late nineteenth century onwards, European wealthy societies gradually became oblivious to the threat of anthropogenic climate change. First, because they grew much more resilient in the face of the vagaries of the heavens. Thanks to railways and agricultural globalization, foodstuffs circulated, breaking the secular cycle of poor harvest, scarcity, and social disturbance. At the same time the forest lost its centrality in the major debates over property, the role of the state, and the overall balance of the economy. Consequently, changes in climate lost their charge of anxiety, their political intensity. Scientific theorizing about climate action gradually tapered off. Human history and climate history had been considered in tandem since the modern age; but the second was now no more than the sometimes variable setting in which the first was played out. This mutation occurred later in the empires, where the thesis of climate action fed into relations of domination for a few decades more. This resilience, this now imperturbable sky, lasted for a brief period of time – scarcely a few decades – during which the climate could seem irrelevant to human life. At the end of our inquiry, we have discovered not the 'origin' of ecological 'awareness', but the opposite: the industrial and scientific engineering of a form of apathy in the face of anthropogenic climate change; the genesis of societies that liked to think they had finally averted this threat.

1

Christopher Columbus's True Discovery

By 1492, it was nearly a century since the Spanish and Portuguese had embarked on their enterprise of conquest in the Atlantic. From the pens of its first chroniclers, the 'discovery of America' appears not as the dawn of a new age, but as an additional stage in a process of expansion initiated with the conquest of the Canaries, Madeira, and Porto Santo. These islands off the coast of Africa have their rightful place in the history of the New World: more than a prelude, they were the laboratory for colonization of the Caribbean, the model for a colonization based on sugar, deforestation, forced labour of the indigenous population, and the importation of African slaves.[1]

Reflections on climate change emerged as colonization of the Canaries and Madeira was nearing completion and that of the Caribbean was beginning; when, on contact with a 'new new world', explorers cast a retrospective glance at what had already been accomplished by their predecessors further east. They interpreted colonization as a profound transformation of Creation, completely altering the climates of the islands they seized, making them dryer and more temperate. This project of

1 See Felipe Fernández-Armesto, *The Canary Islands after the Conquest: The Making of a Colonial Society in the Early Sixteenth Century*, Oxford: Oxford University Press, 1982 and *Before Columbus: Exploration and Colonization from the Mediterranean to the Atlantic, 1229–1492*, Philadelphia: University of Pennsylvania Press, 1987; Anthony M. Stevens-Arroyo, 'Inter-Atlantic Paradigm: The Failure of Spanish Medieval Colonization of the Canary and Caribbean Islands', *Comparative Studies in Society and History*, vol. 35, no. 3, 1993, pp. 515–43.

climate improvement was advanced by the original conquistadors and, in the first instance, by the most illustrious of them all: Christopher Columbus.

'The Trees Produce Clouds and Rain'

July 1494: Columbus's expedition is navigating along the coast of Jamaica under a clear blue sky. Suddenly, it clouds over and torrents fall on the ships. The expedition is in danger: water gets into the holds, corrupts the provisions, and the stifling heat makes it impossible to conserve food. On several occasions, the fleet has to restock from local tribes. In this critical situation, according to his son Ferdinand, Christopher Columbus made the following observation: 'The sky, the disposition of the air and the weather in these places are the same as in the environs' – that is, 'every day, at the hour of Vespers, a cloud appears with rain lasting an hour, sometimes more, sometimes less.' The cause of this strange phenomenon, according to him, lay in 'the country's big trees.' The hypothesis of a link between forest cover and precipitation was presented as personal. Columbus knew 'from experience' (*per esperienza*), Ferdinand tells us, that things had once been the same 'in the Canaries, in Madeira and in the Azores.' But ever since the trees 'which encumbered [*ingombravano*] them' had been cut down, 'the same quantity of clouds and rain was no longer generated.'[2] The testimony is indirect but may be taken seriously. Ferdinand Columbus was a great man of letters, founder of the Columbine library of Seville. He was also the first biographer of his father, whom he accompanied on his fourth and last voyage. He claimed to have been a direct witness of the phenomenon, extraordinary at the time, of tropical rains, daily deluges as regular as clockwork.[3] Above all, Christopher

2 Fernando Colombo, *Historie del S.D. Fernando Colombo; nelle quali s'ha particolare, & vera relatione della vita, & de' fatti dell' Ammiraglio D. Christoforo Colombo, suo padre*, Venice: Franceschi Sanese, 1571, pp. 117–18.

3 Ibid., p. 113. On other occasions, Christopher Columbus advanced meteorological arguments. For example, in 1492 in Cuba, he observed that 'the air is more temperate than elsewhere, because of the height and beauty of the mountains.' This 'beauty' almost certainly related to the forests. Columbus remarked with astonishment that the steep slopes were continuously wooded, without it being possible to make out naked rock (Christophe Colomb, *Journal de bord, 1492–1493*, Paris: Imprimerie nationale, entries for 14 and 26 November 1492). We also know that he discussed with the Italian humanist

Columbus's reasoning made profound sense: it was a question of reassuring his patrons Ferdinand and Isabella about the habitability of lands situated in the 'torrid zone' – supposedly empty of human beings – of ancient geography.

In 1492, cosmography had only been taught in universities for a few decades. Ptolemy's *Geography* had just been rediscovered, as had the classical works of Strabo and Sacrobosco. These authors popularized two notions: 'climate' and 'zone'. The first pertains to fixism: here climate is a purely geodesic notion, without meteorological content, a simple space of land between two parallels.

The notion of zone is more problematic: while predominantly defined by latitude, it nevertheless connects to the physical features of the place. Zones are characterized by their degree of habitability, which depends on the prevailing heat, cold and humidity there. In the ancient cosmography adopted by Arab scholars and doctors, and then by humanists, three of the five zones were deemed uninhabitable: the Arctic and the Antarctic, on account of the cold, and the 'inflamed' zone[4] lying between the tropics of Cancer and Capricorn. This torrid zone was not only deemed uninhabitable, but also, on Aristotle's authority, regarded as impassable, thus separating the two temperate zones to the north and the south. In sum, the ecumene – the known world – formed but a narrow strip relative to the size of the globe, comprising Europe, North Africa and a very shrunken Asia.[5]

The second half of the fifteenth century was marked by a progressive recasting of geographical knowledge, which now envisaged an ecumene encompassing the whole planet. The stakes were scientific, commercial, and theological: the complete opening up of the world thanks to Christian maritime technology made it possible to envisage humanity's universal redemption. Even before the New World intruded into European

and chronicler of the New World, Pietro Martire d'Anghiera, the power and extraordinary width of the rivers he discovered in Central America. Cf. d'Anghiera, *De Orbe Novo*, Paris: Ernest Leroux, 1907, p. 210.

4 Antoine de Pinet, *L'Histoire du monde de Pline*, Cologne: Jacob Stoer, 1608, vol. 1, p. 63.

5 On the habitability of the globe, see Jean-Marc Besse, *Les Grandeurs de la Terre. Aspects du savoir géographique à la Renaissance*, Paris: ENS Éditions, 2003, pp. 66–75; John M. Headley, 'The Sixteenth-Century Venetian Celebration of the Earth's Total Habitability: The Issue of the Fully Habitable World for Renaissance Europe', *Journal of World History*, vol. 8, no. 1, 1997, pp. 1–27; Craig Martin, 'Experience of the New World and Aristotelian Revisions of the Earth's Climates during the Renaissance', *History of Meteorology*, vol. 3, 2006, pp. 1–15.

cosmology, the torrid zone's habitability had become a humanist emblem of the supersession of antiquity's stock of knowledge. Columbus was a contemporary of this first cosmographic revolution. Around 1482, he voyaged as far as São Jorge de Mina, a Portuguese fortress on the coast of Guinea, in Akan territory, below the Tropic of Cancer and hence in the heart of the torrid zone. After this voyage, in the margin of a cosmographical treatise, he noted: 'the torrid zone is not uninhabitable; the Portuguese are sailing it today. It is even highly populated. The fort of the mine we saw stands on the line of the equator.'[6]

Still, despite the Portuguese trading posts in Africa, the habitability of the torrid zone remained questionable for late-fifteenth-century intellectuals. For example, in 1507 Martin Waldseemüller's famous map of the world indicated in a side bar that 'the torrid zone is not desolate' but only 'habitable with difficulty'. Christopher Columbus therefore urgently needed to demonstrate not only the habitability of the tropics, but also their hospitality. And that is why he noted in his diary a whole host of observations on the enchanting climate of the places he discovered. The temperature in Cuba was described as that of a 'May night in Andalusia': 'I certify to Your Highnesses [Ferdinand and Isabella of Spain] that there does not exist under the Sun [a place] superior in fertility and the moderation of heat and cold, and for the abundance of clean, healthy water, unlike that of the rivers of Guinea, which are pestilential.'[7]

If, as is likely, Christopher Columbus actually reflected on climate change, two questions remain: why did he concern himself with it, and where did the idea (presented by his son as being his own) come from?

To answer the first question, we must start with Columbus's general project. Dreaming of crusades, the recapture of Jerusalem and universal conversion, in practice he was also a merchant.[8] His voyages were financed not only by the Spanish monarchy, but also by the bankers of Seville, to whom he had to justify his expeditions by demonstrating the commercial opportunities they would offer. In the late fifteenth century two models were possible: the Portuguese trading posts established on the

6 Pierre d'Ailly, *Ymago Mundi*, Paris: Librairie orientale et continentale, 1930, p. 197. See Peter Hair, 'Was Columbus' First Very Long Voyage a Voyage from Guinea?', *History in Africa*, vol. 22, 1995, pp. 223–37.

7 Colomb, *Journal de bord*, 23 October and 27 November 1492.

8 Denis Crouzet, *Christophe Colomb. Héraut de l'Apocalypse*, Paris: Payot, 2006 and Jérôme Baschet, 'Le *Journal de bord* de Christophe Colomb', in Patrick Boucheron, ed., *Une histoire du monde au XVe siècle*, Paris: Fayard, 2009, pp. 582–7.

coasts of West Africa, and the settlement colonies – producing timber and sugar – of Madeira and the Canaries. Prior to his initial voyage, Columbus favoured the first option. Like the merchants of Lisbon enriched by trade in gold, ivory and slaves on the coast of Guinea, Columbus intended to join in the trade networks of the Great Khan, whom he hoped to meet – and convert – across the Atlantic.[9] In accordance with the cosmographical doctrines of the age, he set out not only westward but also *southward*, for he thought the tropical territories still to be discovered would be similar to the posts of Guinea: the same latitudinal zone, the same astral influences, the same people still in their infancy (who could therefore be reduced to slavery), and, above all, the same natural riches. Likewise for gold: since it was found in quantity on the coasts of Guinea and since, according to alchemical doctrine, it resulted from a process of purification peculiar to the torrid zone, this metal must also be plentiful on the same latitude on the other side of the Atlantic Ocean.[10]

But when, in autumn 1492, Columbus stumbled on the Caribbean islands, which he continued to mistake for Japan, he did not find the wealthy merchant towns covered in gold that Marco Polo, his guide, had described. He then fell back on a different interpretation: the islands of the Caribbean were now described in the light of the archipelagos of Madeira and the Canaries, flourishing economically at the time thanks to wood and sugar. Hence the ubiquity in his diary of a merchant's view of tropical nature. Of Cuba he extolled the trees, 'enormous and straight' like 'the masts of the largest Spanish ships'.[11] Everything was set fair for the merchants of Seville and the Spanish Crown to haul in a fabulous booty, not of gold, ivory and spices, but of timber and sugar.[12]

Let us now return to the coasts of Jamaica, under the driving rain of July 1494. Columbus described the island emphatically as 'the most beautiful he had ever come upon', with a 'vast, convenient port' and 'excellent

9 William and Carla Philips, *The World of Columbus*, Cambridge: Cambridge University Press, 1992 and William Philips, 'Africa and the Atlantic Islands Meet the Garden of Eden: Christopher Columbus's View of America', *Journal of World History*, vol. 3, no. 2, 1992, pp. 149–64.

10 Nicolas Wey Gómez, *The Tropics of Empire: Why Columbus Sailed South to the Indies*, Cambridge, MA: MIT Press, 2008; Peter Hair, 'Columbus from Guinea to America', *History in Africa*, vol. 17, 1990, pp. 113–29.

11 Colomb, *Journal de bord*, 25 November 1492.

12 Colombo, *Historie del S.D. Fernando Colombo*, p. 61.

shelters'.[13] But the torrents of water that fell on his fleet dampened this Edenic image. And it is at this key point in the text that the possibility, the certainty even, of a change in climate is invoked. What had to be understood was that the exploitation of wood – in itself highly profitable – would free the islands of the forests that '*encumber*' them;[14] it would transform their climate, for (according to Columbus) the trees '*produce* clouds and rain'. Finally, cutting down the forests would free up the space required for sugar cultivation. In April 1494, three months prior to the experience of the torrential rains, the Genoese seafarer had 'marvelled' at the extreme fertility of the soils of Hispaniola (Santo Domingo) and the remarkable success of the cane plants he had brought from the Canaries.[15] Presenting the torrential rain as a mere meteorological artefact made it possible to normalize Caribbean nature, to fit it into the Canaries model, and thus remove the climatic obstacle to colonization.

Second question: where did Columbus get the idea – at first sight strange – that the forest 'produces' rain? The seafarer's climate theory stemmed from his direct experience (Fernando speaks of *esperienza*) of a major 'ecological shock'. The islands of Madeira and Porto Santo – unpopulated prior to the arrival of the Portuguese in 1419 – underwent abrupt environmental changes in the space of a few decades. In the 1450s, Portuguese, Flemish and Italian capital, as well as the African slave trade, converged on Madeira. Over the next decades, the 'island of wood', entirely covered with huge trees, became the first global centre of sugar production. Cooking the cane juice required massive quantities of wood: at the peak of production around 1505, it is estimated that the refineries of Funchal (the capital of Madeira) consumed 500 hectares of forest per annum. Added to this were slash-and-burn agriculture and exploitation of timber. Around 1510, one-third of the island's surface area had been cleared, especially along the coasts. For want of fuel, and with soil erosion, sugar production

13 Ibid., p. 110.

14 Let us note that the English and French translations of Columbus's diary do not convey the pejorative tone of this verb. This doubtless explains Richard Grove's attribution to Columbus of a negative view of deforestation, actually making him a 'precursor' of environmental alarm. Cf. Grove, 'Conserving Eden: The (European) East India Companies and Their Environmental Policies on St. Helena, Mauritius and in Western India, 1660 to 1854', *Comparative Studies in Society and History*, vol. 35, no. 2, 1993, pp. 318–51; *Green Imperialism*, p. 31.

15 Colombo, *Historie del S.D. Fernando Colombo*, p. 106.

collapsed, falling from 2,500 tons to 300 in 1530. And when, in the second half of the sixteenth century, Madeira's merchants converted to viticulture, they were forced to import the wood for their casks from New England.[16]

As a Genoese merchant based in Lisbon in the 1470s, Columbus lived through this sugar boom. A trade dispute involving him indicates that he travelled to Madeira on several occasions to acquire cargos of sugar.[17] Most important, in 1478 he married Felipa Moniz, daughter of the conqueror of Porto Santo, Bartolomeu Perestrelo, a Lombard knight who in 1418 had taken part in the very first expedition to the archipelago. After his marriage, Columbus settled for a while (1478–81) in Porto Santo, which was governed at the time by his brother-in-law, the son of Bartolomeu Perestrelo, then in Funchal. As at once a sugar merchant, the son-in-law of the conqueror of Santo Porto, and the brother-in-law of the governor, Columbus was ideally placed to understand the ecological shock generated by colonization.

Several accounts stress the abruptness of the changes undergone by the insular environments. For example, according to a Portuguese chronicle, Bartolomeu Perestrelo, landing on the island of Santo Porto in 1418, introduced rabbits which so damaged the vegetation that the colonists were supposedly compelled to flee the island.[18] Or again, according to the Venetian merchant Cadamosto, in 1419 the Portuguese, aspiring to cultivate Madeira, decided to set fire to its vast forests. The ensuing conflagration 'obliged the governor, in order to escape the fury of the flames, to retreat into the sea, where he remained two days and two nights up to his neck in the water without drinking or eating. . . . By this means they *relieved* [our emphasis] the island of part of its wood, clearing the land for ploughing.'[19]

---

16 Alfred Crosby, *Ecological Imperialism: The Biological Expansion of Europe, 900–1900*, Cambridge: Cambridge University Press, 1986, pp. 70–103; Jason Moore, 'Madeira, Sugar, and the Conquest of Nature in the "First" Sixteenth Century', *Review: A Journal of the Fernand Braudel Center*, vol. XXXIII, no. 1, pp. 345–91.

17 Jacques Heers, *Christophe Colomb*, Paris: Hachette, 1991, pp. 61–4.

18 Gomes Eanes de Zurara, *Chronique de Guinée* (1453), Paris: Chandeigne, 1994, p. 235.

19 Alvise Cadamostro, *Relation des voyages de la côte occidentale d'Afrique* [1455], Paris: Le Roux, 1895, p. 27.

The Sacred Tree of El Hierro

A third account circulating among Spanish conquistadors, merchants, settlers, and scholars must have fed into Columbus's thinking on climate. It concerns El Hierro, the most westerly island in the Canaries. At the time, it symbolized the confines of the known world. Ptolemy had referred to it as 'meridian' and, in his on-board diary, Columbus expressed the distance travelled westwards starting from its location. Zero point for Spanish seafarers and cosmographers, El Hierro was also famous for sheltering a prodigious tree – the sacred tree (*el árbol santo*) – that was said to have the property of attracting clouds and condensing them into rain. This tree, uniquely drawing water from clouds, supposedly supplied all the island's inhabitants, as well as their livestock, with water.

The first description of El Hierro goes back to its attempted capture by Norman knights in 1402. Pierre Bontier and Jean Le Verrier, two Franciscan brothers who accompanied them, highlighted the lack of water

'The Sacred Tree of El Hierro', from Théodore de Bry, *Peregrinationes in Indiam orientalem et Indiam occidentalem*, Frankfurt, 1593, vol. 6.

on the island and the presence of extraordinary trees at its summit. They described 'a high, rather flat country, full of bocage, pines and bay trees . . . at the country's highest point are trees forever dripping down a beautiful clear water which falls in ditches near the trees, the best drinking water to be found.'[20] Subsequently, the island was landed on several times, but attempts to capture it from the Guanches (a people of Berber origin) failed for lack of water. Legend had it that the natives knew a celestial source, a prodigious tree (the *garoé* in the Guanche language) that attracted clouds and turned them into rain. The secret was well guarded until, the story goes, a Guanche woman enamoured of a Spanish soldier revealed it to the conquistadors.[21]

The trees referred to by the Franciscan brothers were transformed into one miraculous tree in the chronicles of the fifteenth and sixteenth centuries. The marvel is recounted by numerous humanists, priests and travellers, including Gonzalo Fernández de Oviedo (to whom we shall return) and the Dominican Bartolomé de Las Casas.[22] In his famous *Historia de las Indias*, Las Casas refers to a 'small cloud' permanently perched at the top of the tree and causing 'its leaves to drip'. According to him this was a 'patent miracle, with no apparent natural cause'.[23]

Columbus, who resupplied all his expeditions in the Canaries (especially in Gomera, very close to El Hierro), is bound to have considered the resources in water offered by the various islands. It is therefore very likely that he was aware of this wonder, famous in the late fifteenth century.[24] And, in July 1494, off the coast of Jamaica, he conferred a new, imperial meaning on it: left to itself, the vegetation had proliferated to

20 Jean de Béthencourt, *Le Canarien. Livre de la conquête et conversion des Canaries*, Rouen: Métérie, 1874, p. 117.

21 Legend recounted by Girolamo Benzoni, *Historia del mondo nuovo* (Venice, 1565), Caracas: Academia nacional de la Historia, 1967, p. 275 and Théodore de Bry, *Tesauro de los viajes a las Indias Occidentales y Orientales*, vol. 6, Liège, 1597, p. 28.

22 D'Anghiera, *De Orbo Novo*, p. 19; Andrés Bernáldez, *Memorias del reinado de los reyes católicos que escribía el bachiller Andrés Bernáldez* (1471–1513), Madrid: Real Academia de la Historia, 1962, pp. 136–7; Vincenzo Pigafetta, *Premier voyage autour du monde dans l'escadre de Magellan* (1536), Paris: Delagrave, 1888, pp. 34–5.

23 Barolomé de Las Casas, *Historia de las Indias* (1540), Mexico: Fondo de cultura económica 1951, p. 110.

24 It is possible that Columbus would have taken on supplies from El Hierro during his first voyage, if a Portuguese squadron had not waited for him there. Cf. *Journal de bord*, Thursday, 6 September 1492. Pietro Martire d'Anghiera claims that Columbus's second expedition, with a strength of 1,200 men, set out from El Hierro: *De Orbe Novo*, p. 19.

such an extent that the clouds attracted by the trees were drowning these lands which were in escheat. The torrential rains proved non-ownership of the Caribbean. This interpretation would enjoy extraordinary success. Because the Canaries represented an obligatory stopover en route to the New World, and also furnished a number of the earliest colonists, their wonders profoundly shaped the understanding of American nature.[25] They were like a model, a dress rehearsal, for the changes to be effected on a much larger scale.

Slavery in a Temperate Zone

As far as we know, the first testimony of a change in climate in America comes from Gonzalo Fernández de Oviedo, military governor in Santo Domingo and royal chronicler of the Indies. When he was writing, in 1548, the improvement of the American climate was no longer a mere hope, as in Columbus, but an attested, widely shared view. Oviedo, who had discussed matters with both 'learned men' and 'natives', explained that he was now convinced that 'these lands [Hispaniola], set foot upon and inhabited by Christians since 1492, are very much changed and are more so every day . . . their constitution is more temperate and it is not as hot . . . On this, all the Spanish living here have been in agreement for some time'.[26] The cooling was even a topic of conversation among the settlers: 'In all the villages of this island, all this is to be observed and there is much discussion of the fact that the air is becoming purer and more refined.' Hispaniola's climate had cooled to the extent that in winter it was not 'uncomfortable to wear a fur coat as in Castille'. Oviedo claimed that the same shift had occurred wherever the Spanish had settled, including

25 Ships were stocked in provisions there, in water but also in men. In the sixteenth century, more than 10,000 Canary Islanders left to settle on the other side of the Atlantic. It was from Tenerife, las Palmas or Gran Canaria that the techniques (Arabic in origin) of cultivation, irrigation and refinement of sugar were transferred to the West Indies. Cf. James J. Parsons, 'The Migration of Canary Islanders to the Americas: An Unbroken Current since Columbus', *The Americas*, vol. 39, no. 4, 1983, pp. 447–81; Analola Borges, 'Aproximación al estudio de la emigración canaria en América en el siglo XVI', *Anuario de Estudios Atlánticos*, no. 23, 1977, pp. 239–62.

26 Gonzalo Fernández de Oviedo y Valdés, *Historia general y natural de las Indias, islas y tierra-firme del mar océano*, Madrid: Imprenta de la Real Academia de la Historia, 1851, vol. 1, pp. 239–41.

on the mainland – for example, in the town of Santa Maria del Darién (in present-day Colombia), where he resided in the 1510s.[27] And also in Mexico: in the 1560s, the university rector conducted a survey among elderly Indigenous people and they confirmed that the rainy season was much shorter since the arrival of the Spanish.[28]

Oviedo based his climate theory on an experience that was incomparable among the sixteenth-century chroniclers. Unlike Pietro Martire d'Anghiera, the coiner of the phrase 'New World' (*orbe novo*), who never crossed the ocean, Oviedo spent half his life across the Atlantic. His *Historia general y natural de las Indias* promoted an empirical epistemology, based on direct experience and first-hand testimony. Even so, it provided a bookish reference to back up its theory. It is drawn from the *Natural History* of Pliny the Elder, a first-century Roman author ubiquitous in Oviedo's writings.[29] The reference is interesting, for it actually highlights the *discontinuity* between ancient and modern thinking about climate change. In his *Natural History* Pliny noted the possibility of anthropogenic climate change. In Thessaly, he wrote, 'in the environs of Larissa, the desiccation of a lake made the region colder and the olive trees, which grew there formerly, stopped appearing.'[30] Perhaps unaware of this passage, Oviedo cited another from the same work, but relating to *astronomical* climatic change attributable to the Earth's motion or a shift in the Sun's trajectory.

According to Oviedo, meteorological phenomena were a matter of both miracles and physics. They were miraculous because God acted through

27 Ibid.

28 Cervantes de Salazar, *Crónica de Nueva España*, Madrid: Hispanic Society of America, 1914, p. 10.

29 Oviedo liked to present himself as 'the Pliny of the New World'. On Oviedo, see Antonello Gerbi, *Nature in the New World, from Christopher Columbus to Fernández de Oviedo*, Pittsburgh: University of Pittsburgh Press, 2020; Louise Bénat-Tachot, 'Les représentations du monde indigène dans la "Historia general y natural des las Indias" de Gonzalo Fernández de Oviedo y Valdés', thesis, Paris III, 1996; Kathleen Ann Myers, *Fernández de Oviedo's Chronicle of America: A New History for a New World*, Austin: University of Texas Press, 2007; and Nicolás Wey Gómez, 'Memorias de la zona tórrida: el naturalismo clásico y la "tropicalidad" americana en el *Sumario de la natural historia de la Indias* de Gonzalo Fernández de Oviedo (1526)', *Revista de Indias*, vol. 73, no. 259, 2013, pp. 609–32.

30 Pliny the Elder, *Natural History*, XVII, III-5, which borrows this anecdote from Theophrastus, *Recherches sur les plantes*, vol. 3, trans. Suzanne Amigues, Budé, 1993, Book V 14, 2–5. On this point, see Glacken, *Traces on the Rhodian Shore*, pp. 127–30. See also Theophrastus, *On Winds*, trans. Robert Mayhew, Leiden: Brill, 2018, p. 168.

them. For example, he punished the idolatrous Indians with hurricanes. Logically, these disasters 'have ceased since the Holy Sacrament was introduced onto the island'.[31] But Hispaniola's cooling had nothing to do with evangelization. Its causes? Ploughing, the proliferation of livestock 'whose breath dispels vapours', and finally the sugar refineries that burnt such quantities of trees 'that it must be seen to be believed'.[32]

Sixteenth-century Spanish discourses on the climate of the New World are invariably bound up with questions of sovereignty and law. In fact, they focus less on the nature of the places conquered than on the legality of their conquest; they seek not so much to describe exotic forms of nature as to naturalize their subordinate position in Spanish imperial geopolitics.[33] The basis of climate improvement, according to Oviedo, was political: desiccation and cooling were due to 'Spanish sovereignty, which tames (*doma*) and mellows (*aplaca*) these regions and their rigours, just as it tames the Indians and animals inhabiting them.' Climate change reflected, hallowed, and sanctified the transition from one sovereignty to another. If, in the initial stages of the conquest, Hispaniola was hot and humid, this was because it had been possessed 'for so long by savage peoples'; because 'neither trodden nor ploughed . . . the forests grew incessantly'; and because 'its very few paths were like rabbit tracks.'[34] The non-domination of nature, and a relationship to the land akin to that of animals, invalidated Indian claims to sovereignty. Oviedo envisaged his natural history as a providentialist treatise on behalf of the global reign of Charles V. Climate improvement attested to a divine plan for Spanish sovereignty over the New World.

Oviedo's climate theory also formed part of the major controversy over possession of the Indies that culminated in Valladolid in 1550–51. It followed Bartolomé de Las Casas's denunciation of the condition of the Indians in 1542 – a denunciation that led to the scheduled abolition of the *encomienda* system, a form of forced labour which, in the event, would survive. Part of what was at stake in the controversy was climatic. Against Las Casas, the theoretician of Spanish imperialism Juan Ginés de Sepúlveda invoked the Aristotelian theory of slavery to justify the

31 De Oviedo, *Historia general*, p. 168.

32 Ibid., p. 240.

33 Rolena Adorno, *The Polemics of Possession in Spanish American Narrative*, Cambridge, MA: Harvard University Press, 2007; Gómez, 'Memorias de la zona tórrida'.

34 De Oviedo, *Historia general*, p. 240.

Indians' serfdom. Natives could be regarded as 'natural slaves' by virtue of the climate they were born in, just like the Blacks in the torrid margins of the inhabited world.[35] By contrast, Las Casas underscored the temperate character of the West Indies climate: there nature, more perfect than elsewhere, had created free individuals, gentle peoples, and kingdoms capable of self-government.[36]

Allied with Sepúlveda, Oviedo found himself on the wrong side of the controversy. In the past, as an inspector of mines, he had actively organized the Indian slave trade. On several occasions he had challenged Las Casas on the legality of the natives' enslavement, and Las Casas had succeeded in blocking publication of the *Historia general y natural de las Indias*, Oviedo's life's work.[37] In the late 1540s, when he drew the portrait of a nature improved by colonization, Hispaniola was in fact in deep crisis: the gold mines were already exhausted, and the demographic collapse of the Indigenous population was a threat to the plantations. His account of an improvement in climate furnished a defensive argument against the charge of ongoing destruction of the Indies.

It also made it possible to remove a contradiction from Spanish imperialist discourse. Apostles of colonization like Columbus, Sepúlveda or Oviedo described the Caribbean islands as the source of unprecedented riches, as spaces more temperate and fertile than the temperate zone itself. And yet these places had engendered peoples inferior to those of the temperate zone: either cannibalistic barbarians, or gentle, effeminate peoples still in their infancy – at all events, peoples incapable of self-government. This rift contradicted the basic precept of Hippocratic, Aristotelian, and Ptolemaic doctrines, which underscored the relationship between the nature of climatic zones and the 'quality' of the inhabitants.

Oviedo tried to resolve this contradiction on several occasions. In his initial texts, he described the Indians as idolators, polygamists, cannibals and sodomites. Whatever the benignity of the West Indian climate, the inferiority of the Indians was rooted in their bodies: 'just as their skull is

---

35 Juan Ginés de Sepúlveda, *Democrates segundo; o, de las justas causas de la guerra contra los Indios* (1545), Mexico City: Fondo de cultura económica, 1996. Cf. Anthony Pagden, *The Fall of Natural Man: The American Indian and the Origins of Comparative Ethnology*, Cambridge: Cambridge University Press, 1982, chapter 5.

36 See, for example, Bartolomé de Las Casas, *Apologética Historia Sumaria*, chapter 20, which compares Hispaniola with England.

37 It would only appear in full in the mid-nineteenth century. See Myers, *Fernández de Oviedo's Chronicle of America*, pp. 55–7.

thick, their reasoning is bestial and malicious.'[38] But, in the 1540s, and in the context of the Valladolid controversy, this commonplace racism was less in favour. The argument of an improvement in climate afforded an ingenious escape clause: it made it possible to exonerate the parenthesis of slavery (given the pre-colonial climate, the Indians were, indeed, natural slaves), as well as providing, in light of recent progress towards a temperate climate, a natural justification for the gradual abolition of the *encomienda*.[39]

It might be said that the 'discovery of the New World' was also the discovery of a new climate – a malleable climate, which Christian, white, industrious man moulded to his advantage. But, in reality, just as America was obviously not 'discovered' by Christopher Columbus, the idea of a climate responding to human action was not some sudden revelation. If Columbus, Oviedo, and other chroniclers considered it unnecessary to shelter behind a decidedly patchy classical authority on the subject, it was no doubt because the idea did not seem particularly surprising to their contemporaries. In accordance with Aristotelian theories, Renaissance natural scientists held celestial phenomena to be closely linked to terrestrial phenomena. Rain and rivers, for example, were part of a single system of water circulation in the sub-lunar world. In addition, in his treatise on *Meteorology*, Aristotle had already stressed the historicity of environments, with their successive, cyclical mutations: rivers that change course, seas that retreat, marshes that dry up and radically alter the meteorological aspect of places. If, according to Aristotle, the history of nature was too slow and gradual to be perceived, by contrast, a change as radical as colonization might well have led to meteorological changes visible within a human lifetime.

---

38 Quoted in ibid., p. 122. Jean-Frédéric Schaub has clearly shown how racism based on genealogy and rooted in bodies emerged from the mid-fifteenth century in Spain with the persecution of the Jews and converts: see his *Pour une histoire politique de la race*, Paris: Seuil, 2015. According to Jorge Cañizares Esguerra, it was in the following century that the hardening of racial differences occurred, within the Creole elite of Spanish ascendancy, born in America, which had to defend itself against accusations of degeneracy. See Esguerra, 'New World, New Stars: Patriotic Astrology and the Invention of Indian and Creole Bodies in Colonial Spanish America, 1600–1650', *American Historical Review*, vol. 104, no. 1, 1999, pp. 33–68.

39 Kathleen Ann Myers clearly shows how Oviedo's description of Indian customs became more obliging, more precise and more 'ethnographic' as the composition of his *General History* proceeded. Cf. *Fernández de Oviedo's Chronicle of America*, chapter 7.

Likewise, the cosmographical notion of zone left a great deal of room for geographical factors altering habitability, sometimes completely. Not everything was determined by latitudinal position – far from it. For example, sub-Saharan Africa ('Ethiopia') and India certainly fell within the hostile margins of the torrid zone, but geographers imagined that the Nile or Ganges made them habitable. In these enclaves, the encounter of humidity and heat engendered marvels and wonders such as elephants, dragons, hippopotami and crocodiles. Thus, the classical theory of zones was freed from the strictly astronomical notion of climate: the meteorological characteristics of places were conceived as largely deriving from the soil, the contours of the terrain, and, above all, from the water circulating between sky and Earth.[40]

In the tradition of these theories, Spanish chroniclers stressed the geographical determinants of the meteorology of the New World. How otherwise could one explain that a space located in the torrid zone could prove so humid and so fertile? The paradox, Oviedo would say, stemmed from its high mountains and immense rivers, those instruments of God: 'this land is naturally hot but, through divine Providence, it is temperate'.[41] The notion of a change in climate through colonial transformation was part of the continuity of divine Creation. The expression 'New World' also signified that the work of Creation was not finished there, and that it was incumbent on Christians to advance it.

40 In this sense, it seems to us mistaken to regard as 'seminal' the very rare mentions of anthropogenic climate change drawn from the ancient corpus. Cf. Grove, *Green Imperalism*, pp. 20–3. See also J. Neumann, 'Climatic Change as a Topic in Classical Greek and Roman Literature', *Climatic Change*, vol. 7, 1985, pp. 441–54. Aside from two brief passages by Theophrastus – one on Crete, the other on Larissa in Attica – the references given by Neumann – Plato's *Critias*, for example – do not mention meteorological phenomena, but only soil destruction and erosion. See also Seneca, *Natural Questions*, III–XI, where the subject is pedology. In plain language: the link between deforestation, erosion, landslides, disappearance of springs and flooding was well-known to the ancients, but this was not the case with the link between deforestation and climate change.

41 Gonzalo Fernández de Oviedo, *Sumario de la natural historia de las Indias*, Mexico City: Fondo de Cultura Económica, 1950, p. 118.

2
Improving the World?

What was to be done with America? In the seventeenth century this was a formidable challenge, posed in this instance to the political, scientific and economic elites of France and England. How were its soil, resources and environments to be exploited? Which subjects and workers should these distant lands be populated with? And how was sovereignty over immense, largely unknown territories to be displayed? America seemed to present itself as a continent predating civilization, as a past from which to set out to map a new path. But which one?

There were plenty of answers. People sought signs. They sought to read into the development of the climate a sanctification of empire. They dreamt of improving Creation by clearing and cultivating the land, by civilizing nature in the New World. Thinking about climate would be all of these at once: justifying conquest, claiming to restore Adam's empire, posing as the sculptor of a nature to be fashioned by reason, force, and the quest for profit. But a shadow of anxiety soon hung over this will to power.

Colonial Propaganda

The first decade of the seventeenth century saw France trying to get a foothold in Canada. Henry IV supported the expeditions of the soldier and the seafarer, Pierre Dugua de Mons and Samuel de Champlain, who founded the towns of Port-Royal and Quebec. This colonizing momentum led to the publication of accounts in the metropolis: the first *History* of the colony by the lawyer Marc Lescarbot, a companion of Champlain, and the *Report* by Father Biard – the first of a long series of Jesuit accounts that would shape the image of Canada.[1] These writings urged colonization and sought to dispel the black legend of a country described by some as a hostile land prey to Satan.[2]

Climate played an important part in them.[3] The authors first of all stressed that France and Canada, being located on the same latitudes, were subject to the same climate in the cosmographical sense.[4] For Biard this helped justify French sovereignty, which was disputed by England. This sovereignty derived, on the one hand, from the anteriority of the discovery and, on the other, from the correspondence of two territories that lay (as the explorer Jacques Cartier had already pointed out) 'under the climates and parallels [of] the regions and kingdom' of France.[5] This was what Biard characterized as the 'natural fairness' favouring French sovereignty, because 'these lands are parallel to our France and not to England.'[6]

But the argument went beyond the idea of spatial continuity. For with the same climate came the same stars and length of days, along with the same temperatures and the same seasons: the same climate in the physical sense of the term.[7] In Lescarbot, this idea underpinned a different argument justifying colonization. His account not only assimilated French

1 Marc Lescarbot, *Histoire de la Nouvelle-France*, Paris: Chez Jean Milot, 1609; Pierre Biard, 'Relation de la Nouvelle-France . . . faite by P. Pierre Biard', in *Relations des Jésuites dans la Nouvelle-France*, vol. 1 (1616), Quebec: Augustin Coté, 1858.

2 As testified a few years earlier by Henri IV's Fool. *La Response de maistre Guillaume au Soldat François*, [n.p.]: 1605, cited by Éric Thierry, 'Le discours démonologique dans les récits de voyages au Canada et en Acadie au début du XVIIe siècle', in Grégoire Holtz and Thibaut Maus de Rolley, eds, *Voyager avec le diable. Voyages réels, voyages imaginaires et discours démonologiques (XVe–XVIIe siècle)*, Paris: PUPS, 2008, pp. 209–20 (here, p. 212).

3 Jean-Baptiste Fressoz and Fabien Locher, 'L'agir humain sur le climat et la naissance de la climatologie historique, XVIIe–XVIIIe siècle', *Revue d'histoire moderne et contemporaine*, vol. 61, no. 1, 2015, pp. 48–78; Colin Coates and Dagomar Degroot, '"Les bois engendrent les frimas et les gelées": comprendre le climat en Nouvelle-France', *Revue d'histoire de l'Amérique française*, vol. 68, nos 3–4, Winter/Spring 2015, pp. 197–219.

4 Biard, 'Relation de la Nouvelle-France', preface; Lescarbot, *Histoire de la Nouvelle-France*, p. 624.

5 Biard, 'Relation de la Nouvelle-France', pp. 1, 66–7; Jacques Cartier, *Voyages au Canada* (1545), Paris: La Découverte, 1992, p. 160.

6 Biard, 'Relation de la Nouvelle-France', p. 66.

7 Ibid., p. 3.

and Canadian climates, it also implicitly contrasted them with that of equatorial Brazil, scene of the debacle of the colony of 'France Antarctique' in 1550–60.[8] It was influenced by Jean Bodin's theories on the correspondence between the character of places and the mores of peoples.[9] While colonial endeavours had failed in Brazil, they would bear fruit in Canada, Lescarbot maintained, for the ways of the French and Indians were similar (or could become so), like their habitats.

But a massive factor destabilized discourses on the homology of climates either side of the Atlantic: Canada's long, freezing winters.[10] In the seventeenth century, description of this cold weather was a staple of any account of Nouvelle-France.[11] As well as marking a rift with French climates,[12] this winter, 'which is so much discussed in Europe on account of its violence and duration' (wrote the Jesuit Lalemant),[13] was detrimental to the cause of colonization. It was to resolve this tension that the thesis of an anthropogenic transformation of climates was articulated, in the early decades of the seventeenth century, as an apologetic discourse for the agricultural exploitation of Canadian lands.

8 Lescarbot, *Histoire de la Nouvelle-France*, pp. 143–228. See Frank Lestringant, 'Champlain, Lescarbot et la "conférence" des histoires', in *Scritti sulla Nouvelle-France nel Seicento, Quaderni del Seicento francese*, Bari, Adriatica and Paris: Nizet, 1984, pp. 69–88, and Michel Bideaux, 'Le discours expansionniste dans l'Histoire de la Nouvelle-France, de Marc Lescarbot', in Frank Lestringant, ed., *La France-Amérique (XVIe–XVIIIe siècles). Actes du XXXVe colloque international d'études humanistes*, Paris: Honoré Champion, 1998, pp. 167–91.

9 Lestringant, 'Champlain, Lescarbot. . .', and 'Europe et théorie des climats dans la seconde moitié du XVIe siècle', in *La Conscience européenne au XVe et au XVIe siècle*, Paris: Collection de l'École normale supérieure de jeunes filles, 1982, pp. 206–26. See Jean Bodin, *Les Six Livres de la République*, Paris: Jacques Du Puys, 1576, Book V.

10 Brian Brazeau underscores this tension in *Writing a New France, 1604–1632: Empire and Early Modern French Identity*, Farnham: Ashgate, 2009, pp. 23–41.

11 Thus, they pepper the Jesuit *Relations* published in France. See Richard Arès, 'Les relations des jésuites et le climat de la Nouvelle-France', *Mémoires de la société royale du Canada*, Series 4, vol. 8, 1970, pp. 75–91.

12 The discrepancy between Canadian winters and the expectations aroused by the identity of cosmographic climates was stressed in the 1540s in the memoir that Jean Fonteneau, the pilot on Roberval's expedition, wrote on his return to France. See Jean Fonteneau (known as Alfonse de Saintonge), *La Cosmographie avec l'espère et régime du soleil et du nord* (1544), 'Recueil de voyages and documents pour servir à l'histoire de la géographie depuis le XIIIe jusqu'à la fin du XVIe siècle', no. 20, Paris: E. Leroux, 1904, pp. 494–5.

13 Hierosme Lalemant, 'Relation . . . ès années 1662 et 1663 [1664]', in *Relations des Jésuites dans la Nouvelle-France*, vol. 3, Quebec: Augustin Coté, 1858, p. 30 (separately paginated).

For Lescarbot, there were two reasons for the abnormal duration of winters across the Atlantic: the inherently cold character of America, but also the immensity of the Canadian forests, which prevented the sun from heating the Earth.[14] Biard took the same line, evoking the 'infinite forest' stopping solar heat from reaching the ground.[15] However, he maintained, this would change were the land to be inhabited and cultivated. Three years earlier, Champlain expressed the same conviction. 'I believe', he insisted, '[that the snow] remains much longer than it would if the land were ploughed.'[16]

As on Hispaniola, the idea of climatic improvement underwrote a project of settlement colonization with a sedentary, agricultural population.[17] It also implicitly condemned two alternative forms of land use. The first was that of the Indians. Endless forests and large rivers were indicative of the vacant state of spaces they did not know how to exploit. An identical argument is to be found in the discourse asserting British sovereignty over New England.[18] The other model to be conspicuously avoided was an economy based on fishing and the fur trade, which took the form of a dispersed presence in the interior and a seasonal presence on the seaboard. The prospect of climate improvement argued for cultivating Canadian spaces. This agricultural route, in addition to its economic, moral and spiritual values, would civilize the climate and then, in a virtuous circle, more clement winters would enable better harvests. The homology of climates and mores would strengthen French sovereignty on these lands.

This vision was also profoundly religious: climate changes generated by agricultural colonization (which was also a work of conversion) showed that such undertakings were hallowed by God. The climate would improve: a further sign of France's legitimacy in taking control of these lands in the name of the king and the Christian faith. The first accounts of concrete

---

14 Lescarbot, *Histoire de la Nouvelle-France*, pp. 624–5.

15 Biard, 'Relation de la Nouvelle-France', pp. 5–6. See also his letter of 31 January 1612 to Claude Aquaviva, Superior General of the Society of Jesus, in Auguste Carayon, ed., *Première Mission des jésuites au Canada. Lettres et documents inédits*, Paris: L'Écureux, 1864, pp. 82–3.

16 Samuel de Champlain, *Les Voyages du sieur de Champlain, xaintongeois, capitaine ordinaire pour le roy*, Paris: Chez Jean Berjon, 1613, pp. 52–3.

17 Lescarbot, *Histoire de la Nouvelle-France*, pp. 156, 228, 482.

18 See William Cronon, *Changes in the Land: Indians, Colonists, and the Ecology of New England*, New York: Hill & Wang, 1983, pp. 54–81.

climate improvement in Canada date from the 1630s, initially from the pen of Jesuit authors.[19]

The argument of climate improvement served above all, and more practically, to recruit colonists. Take, for example, *The Description of North America* published by the entrepreneur Nicolas Denys in 1672, in the wake of the colonial policy promoted by Colbert.[20] Denys spent more than forty years in Canada, at the head of a fishing and trading company operating in Acadia. He also introduced colonists to cultivate the land. If he opted to make himself heard in the metropolis, it was to counter gripes about cold winters that discouraged possible candidates: climate change was a centerpiece of his plea. He compared French and Québecquois winters, attributing the length and harshness of the latter to the accumulation of snow under the immense forest cover. And he continued: 'This can . . . be proved by Quebec, which has two fewer months of Winter than it had before the land was cleared . . . consequently, there is no longer any reason to disparage the country for its great colds and deep snows . . . the new France can produce everything just as well as the old France, but it needs people to work on clearing land.'[21] In short, the colonists' deliberate action had already rectified the climate in some places; and this opened up immense prospects for those willing to settle there.

Similar hopes prevailed at the time in New England.[22] From the 1620s, colonial literature highlighted the possibility of improving its climate

19 In 1633, the Jesuit Le Jeune reported the first observations of a palpable climatic improvement *in situ*: Paul Le Jeune, *Relation de ce qui s'est passé en la Nouvelle-France en l'année 1633*, Paris: S. Cramoisy, 1634, pp. 106–7. Another Jesuit, François Le Mercier, promised in 1665 that the region would become 'a country that can hold its own, as regards fertility of the land and mildness of the climate, before whatever is most mild and most agreeable in Europe', thanks to climate change through deforestation. Quoted by Coates and Degroot in '"Les bois engendrent les frimas et les gelées"', p. 216.

20 Nicolas Denys, *Description géographique et historique des costes de l'Amérique septentrionale*, 2 vols, Paris: Chez Claude Barbin, 1672.

21 Ibid., vol. 2, pp. 8–12 (here, pp. 11–12).

22 Here we rely on Brant Vogel, 'The Letter from Dublin: Climate Change, Colonialism, and the Royal Society in the Seventeenth Century', *Osiris*, vol. 26, no. 1, 2011, pp. 111–28. On the mystery of American climates, see Sam White, 'Unpuzzling American Climate: New World Experience and the Foundations of a New Science', *Isis*, vol. 106, no. 3, 2015, pp. 544–66; Karen Ordahl Kupperman, 'The Puzzle of the American Climate in the Early Colonial World', *American Historical Review*, vol. 87, no. 5, 1982, pp. 1262–89, and 'Climate and Mastery of the Wilderness in Seventeenth-Century New England', in David D. Hall and David G. Allen, eds, *Seventeenth-Century New England: A Conference*

through deforestation, marsh drainage, and agriculture.[23] This discourse partook of the ideology of 'improvement', central in England and which insinuated itself everywhere, from personal morality to agriculture, the army, or London's air. In a world saturated with Providence, how could the efforts of the industrious Protestant not be rewarded by God? Nature was conceived as affording a well-nigh infinite potential for improvement, if only people took the trouble to work it.[24] In America, improvement began with the climate, which was to be purified of its noxious air, dried out and warmed up.[25]

As in France, this discourse was articulated by merchants and entrepreneurs in search of new colonists. In the 1630s, the entrepreneur William Wood asserted that New England's climate had already improved and, he dared say, was even better than that of the mother country – less humid, but not unduly dry.[26] In the colonial literature, climate change was often mentioned in the same chapter as 'seasoning' – the sometimes fatal confrontation of bodies with climatic contrasts – and, at the same time, 'commodious weather' – the ideal weather that prevailed across the Atlantic, ideal at once for European bodies *and* the production of 'commodities'.

From the 1670s onwards, the colony of Virginia became the prime example of a climate that had grown milder, where the dreaded adaptation of bodies to the New World had been alleviated: 'the seasoning was formerly more violent and dangerous here to the English at their first landing', wrote *Speed's Theatre of the Empire of Great Britain*.[27] This was not only to hold out the promise of prosperity, but quite simply to

Held by the Colonial Society of Massachusetts, Boston: Colonial Society of Massachusetts, 1984, pp. 3–38.

23 Richard Whitbourne, *A Discourse and Discovery of New-Found-Land*, London: Felix Kingston for William Barret, 1620, p. 57; Richard Eburne, *A Plaine Path-Way to Plantations*, London: G.P. for Johan Marriott, 1624, p. 22; and John Mason, *A Briefe Discourse of the New-found-land*, Edinburgh: Andro Hart, 1620.

24 Paul Warde, 'The Idea of Improvement 1520–1700', in Richard Hoyle, ed., *Custom, Improvement and the Landscape in Early Modern Britain*, Farnham: Ashgate, 2011, pp. 127–48.

25 Cutting down trees, Richard Whitbourne explained, 'make[s] the country much the hotter winter and summer': *A Discourse and Discovery of New-Found-Land*, p. 57.

26 'Of late the seasons be much altered, the rains coming oftener but more moderately, with lesser thunder and lightnings and sudden gusts of wind': William Wood, *New England's Prospect*, London: Tho. Coates for John Bellamie, 1634.

27 *Speed's Theatre of the Empire of Great Britain*, London, 1676, p. 209, quoted in Vogel, 'The Letter from Dublin'.

reassure applicants for the journey about their chances of surviving.[28] Significantly, the discourse on Virginia's climate ebbed when, following Nathaniel Bacon's Rebellion, settler colonialism was abandoned in favour of a servile workforce, which did not need to be persuaded of anything.[29]

'Cosmical Suspicions'

The thesis of North American climate change spread rapidly in England, from travel literature and apologias for colonization to the scientific world. In the 1660s the latter was organized around the Royal Society and closely connected with colonial trade circles.[30] The colonies recently acquired by England (Ireland, Jamaica, Barbados, New England) were a significant source of income for scientists. Robert Boyle, for example, the founding figure of the Royal Society, owed his fortune to the colonization of Ireland. He was also governor of the New England Company, a member of the Council for Foreign Plantations, and owned shares in the Hudson Bay Company. In addition to this enlightened self-interest, the colonies furnished experimental sites for the Baconian philosophy of mastery over nature and for government steered by political arithmetic. The tight relations between colonial administrators and the Royal Society held out the hope of applying the precepts of empiricism over vast stretches of territory. Guided by experimental philosophy, colonization was to demonstrate its validity and restore Adam's dominion over Creation.[31]

---

28 Virginia was the archetype of a climate improved by the action of Europeans, while another colony, Carolina, was praised for its *naturally* good and fertile climate – a veritable invitation to colonize. See H. Roy Merrens and George D. Terry, 'Dying in Paradise: Malaria, Mortality, and the Perceptual Environment in Colonial South Carolina', *Journal of Southern History*, vol. 50, no. 4, 1984, pp. 533–50. The same type of discourse is found in connection with Georgia in the first half of the eighteenth century: Mart A. Stewart, *What Nature Suffers to Groe: Life, Labor and Landscape on the Georgia Coast, 1680–1920*, Athens: University of Georgia Press, 1996, pp. 35–6.

29 Vogel, 'The Letter from Dublin', p. 119.

30 See Larry Stewart, 'Other Centres of Calculation, or, Where the Royal Society Didn't Count: Commerce, Coffee-Houses and Natural Philosophy in Early Modern London', *British Journal for the History of Science*, vol. 32, no. 2, 1999, pp. 133–53.

31 Sarah Irving, *Natural Science and the Origins of the British Empire*, London: Pickering & Chatto, 2008.

This was the context, merging science, empire and religion, in which Robert Boyle broached the issue of climate change in 1671.[32] In his *Cosmical Suspicions*, a text that stands somewhat apart in his oeuvre, he examined a class of phenomena that eluded the investigations of experimental philosophy and, in particular, mechanistic interpretations: atmospheric conditions, floods, epidemics, the alterations of climate.

To characterize the latter, Boyle mobilized two types of evidence. First of all, colonial promotional literature: he cited passages from Wood's *New England's Prospect* describing the moderation of the seasons resulting from settler activity. Next, direct accounts. Boyle explained that he had himself questioned a 'gentleman' from New England who testified that the climate there had become milder. Above all, he had personally witnessed the reception by King Charles II of the governor of Connecticut, John Winthrop, in the Palace of Whitehall in February 1662. Winthrop had come to swear allegiance and request the grant of a charter for his colony. In fact, the colony had already been created without the assent of the metropolitan authorities and, in the early years of the Restoration, found itself in the dock on account of its members' putative sympathies for Cromwell and the regicides.[33]

During this audience, Charles II apparently questioned Winthrop expressly about the 'air temperature'. It seems that the governor's response, given as a direct quotation, was that the climate had shifted to become much warmer since the settlement of the English. It is impossible to assess the tenor of the exchange, but Boyle's intellectual project, entirely based on the production of reliable evidence involving the aristocratic honour of all participants, lends it undeniable weight.[34] However that may be, the reference to climate change at the highest level is striking. And trusting in

32 Robert Boyle, *Tractatus de cosmicis rerum qualitatibus, cosmicis suspicionibus etc.*, Amsterdam: J. Janssonius Waesberge/Hamburg: G. Schultze, 1671. English edition consulted: 'Cosmical Suspicions', in *The Works of the Honourable Robert Boyle*, London: W. Johnston *et al.*, 1772, vol. 3, pp. 316–25.

33 On John Winthrop (not to be confused with his father of the same name), see Walter W. Woodward, *Prospero's America: John Winthrop, Jr., Alchemy, and the Creation of New England Culture, 1606–1676*, Chapel Hill: Omohundro Institute and University of North Carolina Press, 2013, and Robert C. Black III, *The Younger John Winthrop*, New York and London: Columbia University Press, 1966 (on the charter, see pp. 219–31). Winthrop was an assiduous astronomer and naturalist and acted as an informant of the Royal Society after his return to New England (see ibid., pp. 307–19).

34 Simon Schaffer and Steven Shapin, *Leviathan and the Air-Pump: Hobbes, Boyle, and the Experimental Life*, Princeton: Princeton University Press, 1985.

the reality of the phenomenon, whose intensity and duration would (he wrote) gradually become clear, Boyle opened up space for inquiry into its nature: was it the product of non-human forces, emanating from the stars, or from the Earth's interior? Or was it the result of human action?

This is the question that Henry Nicholson, a former student at Trinity College with an interest in the natural sciences, sought to answer in a letter probably sent to Boyle and published by the Royal Society in 1676.[35] Nicholson was reacting to the idea, seen as commonplace among American colonists, of ongoing climate improvement under the impact of colonization, clearing and sowing. But his argument addressed a different terrain of English colonial expansion: Ireland. The latter, he explained, was less populated and less well-cultivated than it used to be. According to its inhabitants, the island had become warmer and drier. This finding contradicted colonial discourse on the *causes* of change, without challenging its *existence*. Nicholson went on to describe the meteorological observations he had made, and whose initial results he passed on so that they might serve to quantify this development, as Boyle advocated. No sooner were meteorological instruments assuming their modern form than people were already using them to document, even to measure, climate changes.

Nicholson's intervention, which went beyond questions about the scale and causes of change, carried a message that surfaced continually at the time: the interpretation of warming as a providential sign, a blessing on English colonization. His letter appeared at a critical point in the history of Ireland.[36] Cromwell's repression of the 1641 rebellion, combined with famines and epidemics, had resulted in demographic collapse. Thereafter, land confiscation in favour of colonists intensified. Nicholson's allusion to Irish depopulation and – more ambiguously – to a less efficient agricultural exploitation must be understood in the light of these factors. And yet, despite this destruction, the island's colonization emerges from his account as blessed by Providence, via a moral judgement of warming that also applied to the trans-Atlantic colonies.

35 'An Extract of a Letter, etc., from Dublin May the 10th, 1676', in *Philosophical Transactions of the Royal Society of London*, vol. 11, 1676, pp. 647–53. This letter was doubtless addressed to Boyle by a Nicholson keen to elicit his good will and to initiate an exchange. The rest of the letter responds point by point to questions raised in *Cosmical Suspicions* (on magnetism, on animal musk).

36 See John Patrick Montaño, *The Roots of English Colonialism in Ireland*, Cambridge: Cambridge University Press, 2011 and Brian MacCuarta, ed., *Reshaping Ireland, 1550–1700: Colonization and Its Consequences*, Dublin: Four Courts Press, 2011.

The decision to address the letter to Boyle was anything but casual. The philosopher's father, Richard Boyle, had been a key figure in the colonization of Ireland, where he had been lord treasurer and carved out an enormous estate. Robert Boyle was born and possessed vast lands there; and when *Cosmical Suspicions* appeared, his brother Richard occupied the same strategic post of lord treasurer in Ireland. This suggests to what extent the providential sanction of climate change might interest Boyle, both intellectually and personally. Nicholson's intervention also bolstered other influential figures in Ireland and the Royal Society, particularly William Petty, founder member of the Society, champion of the island's 'improvement' through colonization, and himself a major landowner.[37]

In France, scientific interest in the thesis of anthropogenic climate change emerged later. Only in the mid-1740s was it discussed in the Académie des sciences by the natural scientist and physician Duhamel du Monceau.[38] His intervention responded to the recent publication of a book on the colonization of Nouvelle-France, by the Jesuit priest Pierre-François-Xavier de Charlevoix.[39] In it the latter, following many others, advanced the thesis of climate improvement through deforestation.[40] Charlevoix was directly involved in France's Canadian policy, as an envoy of the regent Philippe of Orleans, tasked with assessing the pertinence of a French push westwards.[41] In his book, he cited the arguments of opponents of colonization: the economic balance sheet was in deficit, and 'the climate there is too harsh.'[42] This was not true, Charlevoix hit back: 'the climate of Nouvelle-France will become milder as the country is uncovered.'

Duhamel supported this pro-colonization thesis. Even better, he had a correspondent on the spot in the person of Jean-François Gaultier, a doctor

37 On the English model of the colonization of Ireland, see Montaño, *The Roots of English Colonialism*. On William Petty and Ireland, see Mary Poovey, *A History of the Modern Fact*, Chicago: University of Chicago Press, 1998, pp. 120–38.

38 'Observations botanico-météorologiques faites à Québec par M. Gaultier', *Histoire et Mémoires de l'Académie royale des sciences* for 1746, pp. 88–97.

39 Pierre-François-Xavier de Charlevoix, *Histoire et description générale de la Nouvelle-France*, 3 vols, Paris: Nyon fils, 1744.

40 'There is no retort to experience, which renders the decrease in the cold as the country is uncovered palpable': ibid., vol. 3, p. 168.

41 David M. Hayne, 'Pierre-François-Xavier de Charlevoix', *Biographical Dictionary of Canada*, biographi.ca/index.

42 De Charlevoix, *Histoire et description générale de la Nouvelle-France*, vol. 1, pp. 173–4.

to the king settled in Quebec.[43] Thanks to him, he reported testimony of warming for the year 1745, which was very mild:[44] some 'old inhabitants' had assured the doctor that the wheat harvests were formerly much later, and not as good.[45] This conviction seems to have been shared. At much the same time, the Swedish naturalist Pehr Kalm took down the words of an old man and of priests from Nouvelle-France, who likewise declared themselves convinced that the winters were not as long and/or were less cold than in the past.[46]

Duhamel sought to objectify the phenomenon by using meteorological instruments, of which he was a major promoter.[47] From 1744, he regularly reported to the Académie on the measures taken by Gaultier on the ground.[48] This initial quantitative data caused memory-based evidence to become increasingly discredited. We have an example of this a few years later, when a military engineer officiating in Quebec presented a work to the Académie on the causes of cold in Canada.[49] Reviewing climate change, he stressed that, in order to judge a development, a 'term of comparison' was required, a temporal perspective on things. Yet, he continued, 'This can only be conveyed by [instrumental] observations, as with any phenomenon with degrees of increase and decrease. . . . Facts about objects of this kind cannot be established on the say-so of an old man,' the officer concluded.

43 On Gaultier, see Stéphanie Tésio, 'Climat et médecine au Québec au milieu du XVIIIe siècle', *Scientia Canadensis: Revue canadienne d'histoire des sciences, des techniques et de la médecine*, vol. 31, nos 1–2, 2008, pp. 155–65, and Slonosky, *Climate in the Age of Empire*, chapter 2.

44 'Observations botanico-météorologiques faites à Québec par M. Gaultier', pp. 88–97.

45 'Journal des observations &c. de Mr. Gauthier à Kebec', pp. 17, 46. Archives of the Paris Observatory, A.A.7.6, nos 4–6.

46 This point is highlighted by Coates and Degroot, '"Les bois engendrent les frimas et les gelées"', pp. 216–17. On Pehr Kalm and climate change, see Fredrik Albritton Jonsson, 'Climate Change and the Retreat of the Atlantic: The Cameralist Context of Pehr Kalm's Voyage to North America, 1748–1751', *William and Mary Quarterly*, third series, vol. 72, no. 1, 2015, pp. 99–126. Kalm also met Gaultier *in situ*.

47 Jérémy Desarthe, 'Duhamel du Monceau, météorologue', *Revue d'histoire moderne et contemporaine*, vol. 57, no. 3, 2010, pp. 70–91.

48 'Observations botanico-météorologiques faites à Québec par M. Gaultier', for 1744, pp. 135–55; for 1745, pp. 194–229; for 1746, pp. 88–97; for 1747, pp. 466–88.

49 'D'une lettre écrite à M. l'abbé Nollet, le 20 juillet 1765, par M. De Caire, chevalier de l'ordre de Saint-Louis, & capitaine au corps du Génie, sur la cause du froid au Canada', *Mémoires de mathématique et de physique, présentés à l'Académie des sciences, par divers Savans, et lus dans les Assemblées pour 1773*, Paris: Imprimerie royale, 1776, pp. 541–52.

The Sacred Tree and the Global Water Cycle

European conquests in the Atlantic simultaneously planted seeds of modern climate optimism and disquiet. The fear of catastrophe has been largely obscured by the thundering assertions of imperial elites as to the climate improvements underway in America. Yet it was present from the 1660s in the writings of English natural philosophers. To track this anxiety, we must return to our starting-pont: the sacred tree of El Hierro. While apologists of Empire – whether Spanish, British or French – had inverted the meaning of the miracle to present deforestation and the plantation economy as a positive operation on Creation, in English natural philosophy the sacred tree was the starting-point for reflections on a dramatic desiccation caused by man. These reflections were few and far between, but they very soon pointed to the catastrophic impact of human action on the climate – and this on a global scale from the outset.

The fame of the sacred tree stemmed from the cardinal role played by the Canaries in the imaginary of British empiricist philosophers. The trans-Atlantic voyage symbolized a supplanting of the knowledge of antiquity. One of the most famous representations of empirical philosophy, the frontispiece of Francis Bacon's *Novum Organum* (1620), features a ship crossing the pillars of Hercules. The image was plagiarized from a 1606 Spanish treatise on navigation, which naturally mentions the island of iron, El Hierro.[50] Bacon himself discussed the question of the sacred tree in his book.[51] In travel narratives, in books of botany and natural philosophy, references to the wonder abound. English botanists made it a particular species and gave it the highly significant name of 'fountain-tree'.[52]

The Royal Society began to take a closer interest in the tree. In April 1665, it compiled a list of seventy questions on vegetation addressed to

50 Andrés García de Céspedes, *Regimento de Navegación*, Madrid: Juan de la Cuesta, 1606, p. 175. On this frontispiece, see Juan Pimentel, 'The Iberian Vision: Science and Empire in the Framework of a Universal Monarchy, 1500–1800', *Osiris*, no. 15, 2000, pp. 17–30 and Jorge Cañizares Esguerra, 'Iberian Science in the Renaissance: Ignored How Much Longer?', *Perspectives on Science*, vol. 12, no. 1, 2004, pp. 86–124.

51 Francis Bacon, *Novum Organum*, New York: Collier, 1902, p. 278.

52 It is thanks to the testimony of a mysterious 'master Lewis Jackson' that the tree of El Hierro made its entry into the monumental *Theatrum Botanicum* by John Parkinson, herbalist to Charles I. It acquired the Latin name *Arbor aquam fundens* or 'fountain-tree' – a name it retained in several European languages. Cf. *Theatrum Botanicum, or an Herball of Large Extent*, London: Thomas Cotes, 1640, p. 1649.

travellers and colonists. One of the questions concerned the tree on El Hierro and sought to confirm the veracity of the wonder.[53] The following year, when England was afflicted by drought, *Philosophical Transactions*, the organ of the Royal Society, published a quite remarkable observation by Dr Henry Stubbe, a physician in Jamaica. Contrary to the prevailing imperial climatic optimism, Stubbe stressed the *deterioration* of the Caribbean climate due to colonization. He generalized the miracle of the sacred tree, advancing the hypothesis that 'some trees attract rain'. Hence, he concluded, 'if you destroy the forests, you reduce or destroy the rains.' The connotation was markedly negative. Stubbe maintained that in Barbados and Jamaica the rains had been halved.[54] This idea was subsequently taken up by John Evelyn, the great English forestry expert: 'Barbados grows every year more torrid . . . and so in Jamaica . . . the rains are observed to diminish as their plantations extend: the like I could tell you of some parts of England not far from hence.'[55]

As already noted, these anxieties about climate change were expressed from the outset in a *global* perspective. Firstly, because (as Evelyn intimates) the process of desiccation through deforestation could occur anywhere on the globe, wherever forests decline. Secondly, because such deforestation was apprehended through a comprehensive way of thinking about the Earth: the natural theology of the water cycle.

This very powerful intellectual schema is fundamental for understanding modern climate concerns and a brief reminder is in order here.[56] The great issue in natural theology was to illustrate the 'wisdom of the Creator' by demonstrating both the interconnection of natural phenomena and

53 Thomas Birch, *History of the Royal Society*, vol. 2, London: Millar, 1756, p. 38 (session of 19 April 1665).

54 'Observations Made by a Curious and Learned Person, Sailing from England, to the Caribe-Islands', *Philosophical Transactions of the Royal Society of London*, vol. 2, 1666–67, pp. 493–500 (here, p. 497). This climatic critique of colonization is rather astonishing, for the Royal Society was one of the major promoters of the plantation economy. It counted among its members protagonists in the Royal African Company (supplier of slaves to the Caribbean islands) and numerous admirers of the sugar economy. We might venture the hypothesis that this intervention is a veiled accusation against the cane planters settled during Cromwell's interregnum. Criticizing the initial phase of colonization for its impact on the climate made it possible to turn over a new leaf in the history of the Caribbean, to be written this time by royalist settlers guided by science.

55 John Evelyn, *Silva: Historical Account of the Sacredness and Use of Standing Groves*, 5th edn, 1729, p. 309.

56 See Yi-Fu Tuan, *The Hydrological Cycle and the Wisdom of God: A Theme in Geoteleology*, Toronto: University of Toronto Press, 1968.

the marvellous economy in the means that govern the order of the world. This thinking is radically anthropocentric: God organized the world for man. Utility thus serves as a starting-point for the arguments. For example, how to explain that the Creator has devoted so much space to oceans, when man needs solid ground for crops and in order to prosper? Where was his wisdom in creating so many useless seas? asked the great natural theologian, John Ray. The answer was bound up with the global water cycle: 'For if there were but half the sea as now, there would be also but half the quantity of vapours and consequently we could have but half so many rivers as there are now.'[57] God had ensured that the oceans were sufficiently large to humidify the Earth. Another question: what purpose is served by uncultivated, impassable mountains? According to the astronomer Edmund Halley, 'it seems to be the design of the hills that their Ridges being placed thro' the midst of the continents, might serve, as it were, for Alembicks to distil fresh water.'[58] And that is not all: in trapping more water than their soil absorbs, mountains generate springs and rivers that transport alluvium to the plains. In short, English natural theology conceived the circulation of water on a *global* scale, with connections between oceans and continents, between tropical zones and temperate zones, between mountains and plains, between rain and soil fertility. Water evaporates from the seas and penetrates the atmosphere, winds push it towards continents, it spreads over the ground in the form of rain, condenses on the mountains, and returns to the ocean via springs and rivers.

This is the powerful comprehensive schema into which the sacred tree was fitted. John Ray viewed it as a key element: the marvel of El Hierro was no mere curiosity, but revealed a much more general phenomenon connected with vegetation, to which fell the role of 'distilling' or condensing the water vapour circulating in the atmosphere.[59] 'We owe part of our Rain, Springs, Rivers [and conveniences of life] to the operation of distillation . . . without which the Earth would scarce be habitable.'[60] Ray adds: 'I have read some of the Philosophers who imagined the Earth

57 John Ray, *Wisdom of God Manifested in the Work of the Creation* [1701], Glasgow: Robert Urie, 1744, p. 87.

58 Edmund Halley, *Miscellanea Curiosa*, London: R. Smith, 1708, p. 11.

59 Ray, *Wisdom of God*, p. 178.

60 John Ray, *Miscellaneous Discourses Concerning the Dissolution and Changes of the World*, London: Samuel Smith, 1692, p. 94.

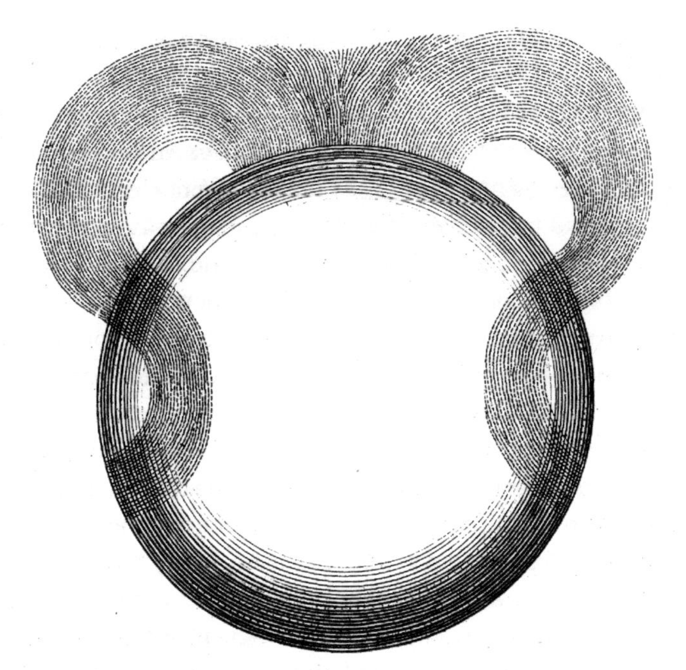

Thomas Burnet, *Sacred Theory of the Earth*, 1690. The author explains how the Earth, smooth and mountainless at the time of Eden, was irrigated thanks to a water cycle linking the tropics to temperate climates.

to be a great Animal . . . and now methinks, if Doctrine be true, we have found out the Circulation of its Blood.'[61] We cannot overestimate the importance of the sacred tree for climate history and, beyond, for the idea of a planet that is reactive and fragile in the face of human action. From the late seventeenth century, it paved the way for the disconcerting hypothesis that man could upset the global water circuit and disturb the providential equilibrium of the Earth, that 'great Animal'.

In 1957, the American physicists Roger Revelle and Hans Suess described global warming as 'a large-scale geophysical experiment'. This phrase, which has become one of the most famous in twentieth-century science, concluded an article crucial for the theory of climate change. The authors demonstrated that with the oceans absorbing only a limited amount of CO_2, the latter was bound to accumulate in the atmosphere. Their concluding phrase contained no irony: Revelle and Suess coldly considered the carbonization of the atmosphere as a scientific experiment

61 Ibid.

that should be closely followed, because it 'could not . . . be reproduced in the future'.[62]

Already in the seventeenth century, scholarly and imperial elites in France and England envisaged the colonization of America as a comprehensive experiment, a massive transformation of Creation which they had both to guide and to observe. This view of the Earth as an experimental object emerges just as clearly from work carried out at the same time on the physiology of plants. In 1699, the English naturalist John Woodward published a memoir that quantitatively studied – thanks to glass containers and a pair of scales – the hydric exchanges between plants and their environment. He described the process of evapotranspiration, the pumping of water into the ground and its return to the atmosphere. It emerged that these exhalations were considerable, varying according to plants and circumstances, but often more than 150 times greater than the weight gain of plants. Each tree, each leaf, each twig becomes like a little source of vapour maintaining humidity in the atmosphere. The most striking thing in the text is the connection Woodward made between the small glass recipients he handled in London and what was transpiring on the scale of the American continent. For, he claimed, 'this so continual an emission of waters in so great plenty from the parts of plants, affords us a manifest reason why countries that abound with trees . . .have more frequent rains'. As regards America, he added: 'The great moisture in the air was a mighty inconvenience to those who first settled in America; which at the time was much overgrown with woods and groves. But as these were burnt and destroyed, to make way for Habitation and Culture of the Earth, the Air mended, changing into a temper much more dry and serene than before.'[63] Woodward thought he had proved the thesis of climate change experimentally. The very particular relationship to the world of Revelle, Suess and Cold War physicists in general, and the detachment with which they could regard climate warming as a 'geophysical experiment', is definitely the legacy of a long history. From its earliest steps, Western science envisaged the planet as one of its instruments, and man's active impact on it as an opportunity for the most formidable of experiments.

62 Weart, *The Discovery of Global Warming*, p. 30.

63 John Woodward, 'Some Thoughts and Experiments Concerning Vegetation', *Philosophical Transactions of the Royal Society*, vol. 21, 1699, pp. 193–227.

3

The Climate of History

For the people living through it, the eighteenth century was an anthropocene: for them, the temporalities of the Earth and of society were one and the same. Human climate action was a modality of this intersection, an index of this temporal concordance.

The naturalist Buffon, one of the most famous scientists of the age, was the herald of a history of the Earth that cast humanity as a planetary force capable of moulding climates. His vision also projected a gauge of civilization: peoples could be hierarchically ranked, he suggested, according to their active impact on the climate, deliberate or involuntary, beneficent or deleterious. Thus, a new divide between 'them' and 'us' emerged at the dawn of climate change.

Buffon's voice was not an isolated one. Throughout eighteenth-century Europe, historiography was inseparably bound up with climate theories, some of which were organized around an anthropogenic change in climates. In this way, people sought to explain the ebb and flow of artistic greatness; to shed light on the origin and history of European nations and monarchies. And, at a time when the history of 'peoples' (Romans, Celts, Germans, etc.) was beginning to be written, the air was full of questions about the interconnected trajectories of populations and climates, which mutually and continuously shaped one another.

In an article entitled 'The Climate of History',[1] the historian Dipesh Chakrabarty offered what is probably the most influential analysis of the

1 Dipesh Chakrabarty, 'Le climat de l'histoire: quatre thèses', *Revue internationale des*

impact of contemporary climate change on our view and our writing of history. He explained that the great divide between 'history of societies' and 'history of nature' was constructed in the modern era, taking for granted the limited impact of human action on natural structures. Humanity having recently become aware of its climatic and even geological power, it had become imperative to rethink and supersede this outmoded rift. The discipline of history was thus interrogated starting from a new 'narrative of origins', based on the great divide between the time of nature and the time of societies – a divide from which we have scarcely begun to free ourselves. But, what if, on the contrary, the idea of a concordance of temporalities was rooted in the cultural framework of modernity?

Why Did the Romans Decline?

The importance of climate for eighteenth-century historians stemmed, in the first instance, from something self-evident, a sort of commonplace in the republic of letters and beyond: the idea that it 'influences' bodies, intellectual output, customs and political institutions. This conception is an inheritance from a very long history, which has been closely studied from the Middles Ages to the modern period.[2] But the notion of humanity's *active impact* on the climate really gained purchase in historical reflection with the appearance in 1719 of a highly influential book by Jean-Baptiste Dubos, on poetry and painting between antiquity and the present time.[3]

livres et des idées, no. 15, 2010, pp. 22–31; Christophe Bonneuil and Jean-Baptiste Fressoz, *L'Événement anthropocène. La Terre, l'histoire et nous*, Paris: Seuil, 2016.

2 The bibliography here is immense. We may cite Sara Miglietti and John Morgan, eds, *Governing the Environment in the Early Modern World*; Sara Miglietti, ed., 'Climates Past and Present: Perspectives from Early Modern France', special issue of *Modern Language Notes (French Issue)*, no. 132, 2017; on the (neo)Hippocratic tradition, David Cantor, ed., *Reinventing Hippocrates*, Aldershot: Ashgate, 2002; Golinski, *British Weather and the Climate of Enlightenment*, pp. 140–50; Ehrard, *L'Idée de nature en France*.

3 Jean-Baptiste Dubos, *Réflexions critiques sur la poésie et sur la peinture*, 2 vols, Paris: Jean Mariette, 1719. On Dubos, see the old but useful work by Alfred Lombard, *L'Abbé Du Bos, un initiateur de la pensée moderne (1670–1742)*, Paris: Hachette, 1913; Jacques Darriulat, 'Jean-Baptiste Du Bos. Réflexions critiques sur la poésie et sur la peinture [1719]', online at jdarriulat.net/Auteurs/Dubos; Dominique Désirat, 'La théorie des climats appliquée aux beaux-arts chez l'abbé Dubos', in Alain Montandon, ed., *Le Même et l'Autre. Regards européens*, Clermont-Ferrand: Association des publications de la Faculté des lettres et sciences humaines de Clermont-Ferrand, 1997, pp. 119–29; Theodore S.

As it took shape in the late seventeenth century in the 'Quarrel of the Ancients and the Moderns', aesthetic reflection also employed climatic arguments.[4] Used by both camps, the notion of climatic influence was, however, double-edged. Orthogonal to the idea of progress dear to supporters of the moderns, it was also problematic for defenders of the ancients: via the climate, it tended to make the mediocre artistic reputation of eighteenth-century Italy and Greece rebound on antiquity.

Such was the challenge that Dubos wanted to take up: to explain how the 'great centuries' – the historical periods marked by formidable artistic fertility – were able to exist under climates unfavourable to the arts today. Before him, this enquiry had given rise to composite theses: the spatial disparity of the arts was due to physical – climatic – causes, their temporal disparity to moral causes, such as education and institutions.[5]

Dubos's explanation, by contrast, closely combined natural habitat and social existence, making human action on the air and temperature the key to the mystery.[6] This went for Rome: Dubos describes the 'major changes in the air' of the city, occasioned by events (destruction of the city's sewers, abandonment of crops, drainage of marshes, expansion of alum mines) which were themselves induced by the fall of the Empire.[7] Over and above the character of the air, according to Dubos, the temperature of Italy had changed: 'the climate' (in the sense of region), he contended, 'is less cold there than it was in the time of the first Caesars'.[8] He argued on the basis of his reading of classical literature – Juvenal, Horace, the *Annals of Rome* – where the freezing-over of the Tiber, for example, is frequently recorded.

There was no need to tie oneself into knots to explain the gulf between the artists of antiquity and those of present-day Italy: the climate quite simply declined along with the mores of the inhabitants, causing genius to ebb. For Dubos, this was a way of putting the ancients and the moderns

Feldman, 'The Ancient Climate in the Eighteenth and Early Nineteenth Century', in Michael Shortland, ed., *Science and Nature: Essays in the History of the Environmental Sciences*, Oxford: BSHS, 1993, pp. 23–40 (esp. pp. 28–30).

4 Michael Cardy, 'Discussion of the Theory of Climate in the Querelle des Anciens et des Modernes', *Studies on Voltaire*, no. 163, 1976, pp. 73–88.

5 See Abbé Bouhours, *Les Entretiens d'Ariste et d'Eugène*, Amsterdam: J. Lejeune, 1671.

6 On Dubos and thinking about climate change, see also Fleming, *Historical Perspectives on Climate Change*, pp. 12–16 and Golinski, *British Weather and the Climate of Enlightenment*, pp. 174–6.

7 Dubos, *Réflexions critiques*, vol. 2, pp. 263–8.

8 Ibid., p. 268.

on the same footing, denying progress in the arts while asserting the historical relativity of good taste.

Dubos's argument is pervaded by an empiricist materialism[9] (he was an admirer of Locke), but also resonates with the discreet echo of theses on American change. The region of Rome had become warmer, he noted, '*although* the country was more populated and better cultivated than it is at present.'[10] His *Réflexions* rapidly won a vast audience, despite their controversial reception,[11] and propagated the notion of anthropogenic climate change throughout the Europe of the republic of letters. Also revealed here is a crucial sub-text of modern thinking about climates: the burning question of the decline of the great ancient civilizations construed in the light of the character of their natural habitats.

We know that Dubos was one of Montesquieu's main sources on climate.[12] The latter read the *Réflexions* before his journey to Italy of 1728–29, and its ideas would permeate his view of the Roman countryside. In his travel notes, Montesquieu highlighted the scarcity of agriculture in the region: 'So here we have a desert!' he wrote, continuing: 'The want of crops produces bad air and bad air has since prevented re-population.'[13] We find no trace of climate change either in *L'Esprit des lois* or in the 'Essay on Causes', a manuscript written in the 1730s, which served as a laboratory for Montesquieu's climatic conceptions.[14] But we have to understand to

9 In *L'Idée de nature en France*, Jean Ehrard makes the rise of materialist philosophy, in its empiricist and sensualist versions, the principal motor of the expansion of climatic rationalities in the eighteenth century.

10 Dubos, *Réflexions critiques*, vol. 2, p. 268 (our emphasis).

11 Roger Mercier, 'La théorie des climats des "Réflexions critiques" à "L'Esprit des lois"', *Revue d'histoire littéraire de la France*, no. 53, 1953, pp. 17–37, 159–74. See, for example, Fréron's acerbic criticisms in *Jugements sur quelques ouvrages nouveaux*, vol. 2, Avignon: Pierre Girou, 1744, pp. 112–23.

12 Mercier, 'La théorie des climats', pp. 159–74; Denis de Casabianca, 'Climats', Dictionnaire Montesquieu en ligne (dictionnaire-montesquieu.ens-lyon); Sheila Mason, 'La physiologie des mœurs selon Montesquieu. Cadre académique et postérité médicale', in Louis Desgraves, ed., *Actes du colloque international . . . pour commémorer le 250e anniversaire de la parution de L'Esprit des lois*, Bordeaux: Académie de Bordeaux, 1999, pp. 387–95; Catherine Volphilhac-Auger, 'Sur quelques sources prétendues du livre XIV de *L'Esprit des lois*. De *L'Essai sur les causes* à *L'Esprit des lois*: la théorie des climats existe-t-elle?' (Article 872, montesquieu.ens-lyon).

13 Quoted in the introduction to Montesquieu, 'Réflexions sur les habitants de Rome', in Montesquieu, Œuvres *complètes*, vol. 9, Oxford: Voltaire Foundation/Naples: Instituto Italiano per gli Studi Filossofici, 2006, pp. 69–82.

14 Montesquieu, 'Essai sur les causes qui peuvent affecter les esprits et les caractères', in ibid., pp. 203–70.

what extent this option derived from a conscious choice. For, at the start of the decade, Montesquieu had no hesitation presenting a memoir to the Academy of Bordeaux in which, in a direct line from Dubos, he studied the development of Roman customs by employing, in addition to moral causes (the habit of bathing, the city's wealth), the argument of a change of air quality generated by the stagnation of water in the ancient ruins.[15] Thus he deliberately chose in *L'Esprit des lois* to describe, *a contrario*, moral causes as the sole motor of the *evolution* of societies and political regimes. In the interim his convictions were refocused around the need for each people to 'make do' with the climate of their natural habitat, regarded as immutable.

The Climatic History of the European Peoples

While absent from *L'Esprit des lois*, until the end of the eighteenth century the thesis of anthropogenic climate change fuelled the growing body of research into the long history of European peoples. This was a new historical genre, featuring not political entities (Empire, Church, or monarchies), but precisely 'peoples' – that is to say, biological entities impacting on territories by proliferation and migration.

Simon Pelloutier, a Huguenot minister and member of the Berlin Academy, was one of the first to take this road. With his research, he aspired to elucidate the origin of all the European peoples. His thesis was the existence of an original Celtic group, issuing from the vicinity of the Danube, which had spread out on a continental scale, before being differentiated over time into Gallic, Germanic, English, Hispanic and Irish nations, and even a Roman one, through contact with the Greeks.[16] Pelloutier described the Celts as a warlike people, hardened by contact with a wild nature and especially a climate much colder than that of eighteenth-century Europe. He adopted the account of climate warming in Dubos, but in order to extend it to the whole continent and corroborate

15 Montesquieu, 'Réflexions sur les habitants de Rome'. The memoir was read to the Bordeaux Academy in December 1732.

16 Simon Pelloutier, *Histoire des Celtes, et plus particulièrement des Gaulois et des Germains*, The Hague: Isaac Beauregard, 1740. On Pelloutier, see Colin Kidd, *British Identities before Nationalism: Ethnicity and Nationhood in the Atlantic World*, Cambridge: Cambridge University Press, 1999, pp. 189–93, 207–9.

it with numerous indices derived from ancient authors: Diodorus of Sicily, Strabo, Virgil, Ovid (freezing of Gallic rivers, presence of bears in the forests of Thrace).[17]

His interpretation linked climate change and the civilizing process. With time, the Celts, putting aside the sword to work the land, had become rooted and settled down, changing the face of the country where they founded the nations of Europe. The clearing and cultivation of land, by allowing the sun's rays to be reflected and the soil to transpire, had warmed the climate. To appreciate Pelloutier's argument, we must go back to the roots of his project: to analyse the trajectory of Europeans from savagery to civilization, by way of a comparison with the case of the autochthonous populations of America.[18] He described Celts and Native Americans as twin peoples, of equal savagery, and possibly of the same origin if (as he thought) the latter descended from migrations out of Eurasia.[19] American climate change, caused by the cultivation of land, served here to think about Europe's original social/natural civilizing process. But it also made it possible to contrast two major historical trajectories: that of Europeans, who since antiquity had embarked on the virtuous cycle of agriculture, sedentary existence, and a milder climate; and that of Native Americans, who had remained on the threshold of acting upon nature. His characterization of the 'European trajectory' gave Pelloutier, a protégé of the King of Prussia, the opportunity for a thinly veiled critique of the historical role of the nobility, whose love of war had delayed the growth of agriculture.[20]

The same kind of enquiry was conducted by Paul-Henri Mallet, a Genevan protégé of the Danish crown, when in 1755 he published his *Introduction à l'histoire de Dannemarc*.[21] In it he described the Nordic peoples' successive migrations from north to south as the engine of the continent's long history. Following Montesquieu, Mallet stressed the free,

17 Pelloutier, *Histoire des Celtes*, pp. 120–4.

18 Simon Pelloutier, 'Lettre de M. P. à M. de B. sur les Celtes', *Bibliothèque germanique*, vol. 28, for 1733, Amsterdam, 1734, pp. 33–51. See also Pelloutier, *Histoire des Celtes*, pp. 239, 295, 457.

19 Simon Pelloutier, 'Dissertation sur un passage des Commentaires de Jules César De Bello Gallico', *Histoire de l'Académie royale de Berlin*, for 1749, Berlin: Haude and Spener, 1751, pp. 491–500.

20 Pelloutier, *Histoire des Celtes*, p. 122.

21 Paul-Henri Mallet, *Introduction à l'histoire de Dannemarc, où l'on traite de la religion, des lois, des mœurs et des usages des anciens Danois*, Copenhagen: Imprimerie des héritiers de Berling, by L.H. Lillie, 1755.

wild character of Northerners and explained it by the climate.[22] But how, then, was the current state of Scandinavian nations and Danish absolutism, for which he was a spokesman, to be understood? First of all, through the action of moral causes. Next, by mobilizing the thesis of a man-made change in climate that had civilized these countries, with reference to the theses of Dubos and Pelloutier, and the writings of Charlevoix on Canada's milder winters.[23]

These histories had a common framework. In the manner of Pelloutier in Berlin and Mallet in Denmark, the Scottish minister Robert Henry penned a *History of Great Britain* in which the Roman conquest had civilized peoples, nature, and climate alike.[24] And the Catholic priest Michael Ignaz Schmidt, famous for having written the first *History of the Germans* which regarded its subject as a single people, deplored the fact that the development of agriculture and milder climates had been checked, in their case, by the great migrations of the early centuries.[25] Relieving demographic pressure, these had caused them to remain arrested at the hunting and nomadic stage, until the Christian religion and the Frankish monarchy liberated them from barbarism and the cold. Even the philosophers of the Scottish Enlightenment (John Millar, Adam Ferguson, and Lord Kames), while highly critical of climatic determinism, granted the climate a certain role in the early phases of societal development. For them, the agricultural stage was precisely the point of escape from climatic determinism, in part because it improved climates.[26]

But, if acting on the climate signals the escape from savagery, the emergence of order may carry the seeds of its own destruction. That, at any rate,

22 Ian Wood, *The Modern Origins of the Early Medieval Ages*, Oxford and New York: Oxford University Press, 2013, pp. 37–45; Thor J. Beck, *Northern Antiquities in French Learning and Literature (1755–1855): A Study in Preromantic Ideas*, New York: Institute of French Studies/Columbia University, vol. 1, 1934, pp. 19–44; Jane Rendall, '"Gothic Feminism" and British Histories, c. 1750–1800', in Geoffrey Cubitt, ed., *Imagining Nations*, Manchester and New York: Manchester University Press, 1998, pp. 57–74.

23 Mallet, *Introduction à l'histoire de Dannemarc*, pp. 252–6.

24 Robert Henry, *The History of Great Britain* [1771], London: Strahan, 1800, vol. 2, p. 287.

25 Michael Ignaz Schmidt, *Geschichte der Deutschen*, Ulm: Stetti, 1778, vol. 1; French translation *Histoire des Allemands*, Liège: Plomteux, 1784. On Schmidt, see Michael O. Printy, 'From Barbarism to Religion: Church History and the Enlightened Narrative in Germany', *German History*, vol. 23, no. 2, 2005, pp. 175–201.

26 Adam Ferguson, *An Essay on the History of Civil Society* [1767], Cambridge: Cambridge University Press, 1992.

is what is suggested by one of the most famous works of history in the late eighteenth century: Edward Gibbon's *History of the Decline and Fall of the Roman Empire*. Gibbon, who was familiar with the climate debates, cited Dubos, Charlevoix and Pelloutier to assert that, in the era of the barbarian invasions, Germany was much more heavily wooded and its climate very harsh. Hence the making of a people that, surpassing the Romans in physical strength, was to bring down the Empire. The trans-Atlantic comparison was still in the background: according to Gibbon, 'Canada, at this day, is an exact picture of ancient Germany.'[27]

Ranking Nations

From discourses on the colonization of North America to narratives of the history of Europe, a fundamental thesis on the links between nature and civilization thus crystallized around the idea of climate action. This thesis was twofold.

First of all, it celebrated the power of humanity, which moulds nature in its own image. Once the barbarous stage has been superseded, once the European peoples abandon war, plunder and migration and devote themselves to working the land, they trigger a virtuous climatic-civilizational circle. For historians, this thesis made it possible to flatter the institutions for which they worked, the absolutist monarchies or the Catholic Church. Moreover, it fitted neatly into the cameralist and populationist discourse of the second half of the eighteenth century, with the virtuous circle amenable to acceleration by princes concerned to increase their population. The argument of climate change was very much bound up with demographic reflections. For example, David Hume invoked climate warming to prove that eighteenth-century Europe was more populous than during antiquity.[28] And Moheau and Montyon, in 1778 in their *Recherches et considérations sur la population de la France*, identified the political springs of demographic growth. Having set out the principles of

27 Edward Gibbon, *The History of the Decline and Fall of the Roman Empire*, London: Strahan, 1776, vol. 1, p. 218.

28 David Hume, 'Of the Populousness of Ancient Nations [1742]', in *Selected Essays*, eds Stephen Copley and Andrew Edgar, Oxford: Oxford University Press, 1996, pp. 223–73. On Hume and climate, see also Golinski, *British Weather and the Climate of Enlightenment*, pp. 177–8.

political economy, they concluded with a climate programme addressed to the king: the monarchy must grasp the whole 'physical order' of its domain, for 'a different climate forms a new species.'[29]

Correlatively, the thesis of anthropogenic climate change, product of the trans-Atlantic cultural matrix, also functioned as an operator of the *ranking* of societies and the trajectories of civilizations. Its emergence marked a redefinition of the criteria of the savage and the civilized, in which religious and moral criteria receded in favour of the capacity to mould nature and, accordingly, to reproduce oneself as a biological entity, a sensitive being, a thinking subject. A product of the upheavals occasioned in Western thought by imperial globalization, the growth of materialist thinking, civil history, philosophy and natural history, climate action became established during the Enlightenment as a new version of the great divide, constantly renegotiated, between 'them' and 'us'.

This dual paradigm linking action on nature and civilization possesses its manifesto: Buffon's Époques *de la nature*, one of the high points of the climatic hubris of the modern age. This text forms part of the enormous *Histoire naturelle* published by the naturalist over several decades. This veritable bestseller of the age of Enlightenment, ubiquitous in the libraries of princes, scholars, and the enlightened bourgeoisie of the late eighteenth century, is peppered with references to anthropogenic climate change.[30] America, and the example of Canada in particular (Chapter 2), were at the heart of these reflections which, as often in Buffon, compared the nature of the 'old' and 'new' worlds.[31]

But, in the Époques, he goes even further by offering a complete history of the terrestrial globe since its creation. He distinguished seven ages. The last was the one that had witnessed the advent of humanity as a global force: 'The whole face of the Earth today carries the imprint of man's power', he wrote.[32] Buffon's seventh age is an Anthropocene, where human action and natural processes intermingle to create a new nature on a planetary scale. Two processes, above all, underpinned this view: the

29 Jean-Baptiste Moheau and Antoine Montyon, *Recherches et considérations sur la population de la France*, Paris: Moutard, 1778, vol. 2, p. 156.

30 Buffon, *Histoire naturelle générale et particulière*, vol. 1, Paris: Imprimerie royale, 1749, p. 242 (forests cause rains); vol. 4, Paris: Imprimerie royale, 1753, p. 397; vol. 12, Paris: Imprimerie royale, 1764, pp. 79–117 (note on 'The Elk and the Reindeer').

31 Buffon, 'The Elk and the Reindeer', p. 92.

32 Buffon, *Histoire naturelle*, supplement, vol. 5 ('Des époques de la nature'), Paris: Imprimerie royale, 1778, p. 237.

domestication, acclimatization and selection of animal species, on the one hand, and human transformation of the climate, on the other. Armed with evidence of the warming of America and Europe, Buffon became the champion of a veritable climate utopia: were they to give up tearing one another apart, civilized nations could rationally transform the planet. For in planting and felling trees judiciously, man could 'alter the influences of the climate he inhabits and fix, as it were, its temperature at the point that suits him'.[33] And because, for Buffon, terrestrial temperatures were what determined the emergence over time of new living species,[34] this geo-engineering would lead to the birth of a new vegetable and animal world fashioned by man.

Buffon's language had a moral, optimistic tonality: it was humanity's duty to assist, embellish and fertilize Nature.[35] But this humanity was split. Arrayed against the European nations were the 'savage little nations of America' and the semi-civilized ones of Africa. 'You can easily judge', said Buffon, 'the minimal value of these men by the minimal impression their hands have made on the soil.' These nations, he complained, 'do nothing but weigh down the globe without relieving the Earth, using everything without renewing anything'.[36] At least the 'Savage' was still weak, whereas 'nations one-quarter polished', like the 'barbarians' of the Middle Ages or the peoples of the Middle East, devastated the Earth and ruined nations. From one continent and one era to the next, the capacity to mould nature and climate placed men on a sliding scale of civilization.

Countering the Encroaching Cold

In the eighteenth century, examinations of climate change were not only available on a historical scale. During the century, a set of naturalist observations perplexed scientists. In sifting through European soils, they unearthed imprints of plants, animal fossils and the bones of mammals

33 Ibid., p. 244.

34 See the update by Jacques Roger, in Buffon, *Les Époques de la nature. Édition critique* [1778], Paris: Éd. du Muséum, 1988, pp. LXX–LXXI. This was not a form of transformism, but a theory of the disappearance/appearance of species starting from inanimate matter, under the influence of temperatures.

35 Thierry Hoquet, 'La théorie des climats dans l'*Histoire naturelle* de Buffon', *Corpus. Revue de philosophie*, no. 34, 1998, pp. 59–90 (esp. pp. 86–90).

36 Buffon, *Histoire naturelle*, supplément, vol. 5, p. 237.

(such as elephants) all known to live in tropical climes. The multiplication of such discoveries throughout Europe with the development of coal mines; the fact that plant fossils varied from region to region much like current flora; finally, the perfect preservation of these imprints, all these led naturalists to reject the initial – diluvial – explanation of their presence. Plant fossils had not been transported by enormous waves but were indigenous; and this implied that local climates had changed drastically over time.[37]

But how were these climate changes to be explained? It was in response to this challenge that Buffon developed the most influential cosmogonic thesis of the age. Under Leibniz's influence, he described the history of the Earth as one of progressive cooling, from the initial state of a melting ball.[38] On its surface, this cooling had resulted in a *climate change on the scale of geological time*, which was the source of waves of emergence and extinction of species.[39] Here Buffon drew on his own experiments to assess the cooling time for spheres of different sizes and compositions, so as to induct a rule that could be transposed to the terrestrial globe.[40] He accordingly described the history of the Earth as one of a progressive cooling extending over 70,000 years. His thesis was that the heat felt on the surface of the ground came almost entirely from inside the globe, not from the Sun. This internal hearth, a residue of the process of planetary formation, was 'the treasure of Nature, the true core of the fire that animates us, as it does all beings'.[41] But this treasure was slowly petering out.

The major thesis of global cooling underpins the successive 'ages of

37 Jean-François-Clément Morand, *Du charbon de terre et de ses mines. Description des arts et métiers*, Paris: Desaint, 1761, p. 168; Romé de L'Isle, *Cristallographie*, Paris: Imprimerie de Monsieur, 1783, vol. 2, p. 600; Jacques L. Bournon, *Essai sur la lithologie des environs de Saint-Étienne-en-Forez*, [n.p.]: 1785, p. 78; Jean-Louis Giraud Soulavie, *Histoire naturelle de la France méridionale*, Paris: Quillau, 1781, vol. 4, pp. 45–8.

38 On the context of the development of Buffon's thesis, and the influence of Leibniz and Dortous de Mairan, see Buffon, *Les Époques de la nature*, pp. xxvii–xxxv, and Jean Gayon, ed., *Buffon 88. Actes du Colloque international pour le bicentenaire de la mort de Buffon (Paris-Montbard-Dijon, 14 au 22 juin 1988)*, Paris: Vrin, 1992.

39 Buffon, 'Premier mémoire. Recherches sur le refroidissement de la Terre et des planètes' and 'Second mémoire. Fondemens des recherches précédentes sur la température des planètes', *Histoire naturelle, générale et particulière, servant de suite à la Théorie de la terre et d'Introduction à l'Histoire des minéraux*, supplément, vol. 2, Paris: Imprimerie royale, 1775, pp. 313–61, 361–77.

40 Buffon, 'Huitième mémoire. Expériences sur la pesanteur du Feu, et sur la durée de l'incandescence', Ibid., pp. 1–36.

41 Buffon, 'Second mémoire', p. 376.

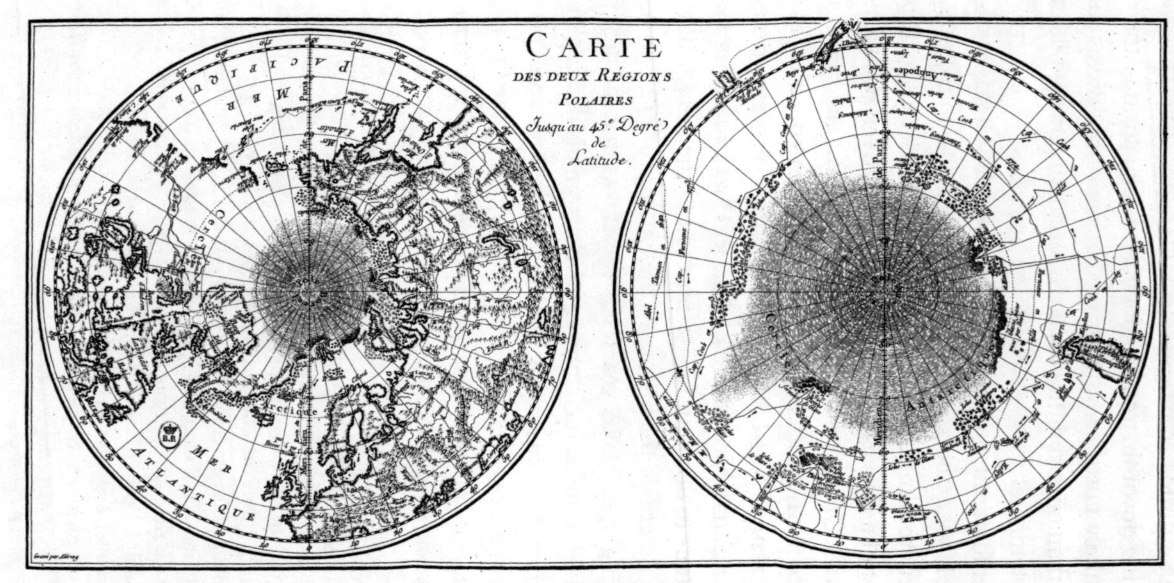

The polar regions were of particular interest to Buffon, who detected signs in them of the encroachment of cold over the globe. 'Map of the two polar regions up to 45 degrees latitude'. Buffon, *Histoire naturelle, générale et particulière*, supplément, Vol. 5 ('Des Époques de la Nature'), Paris: Imprimerie royale, 1778, p. 614.

nature' described by Buffon. By this means he was able to propose a unitary narrative embracing, at one and the same time, the formation and geological evolution of the globe; the history of species; the transformation of climates; and a solution to the mystery of tropical fossil species. The planet was rushing down the road to thermal death. But this end could perhaps be infinitely postponed: 'Nothing seems more difficult, not to say impossible, than opposing the consecutive cooling of the Earth and warming up the temperature of a climate. Yet man can do it and has done it.'[42] Anthropogenic warming and civilization were not a matter of choice: they were the only weapons against the encroaching cold.

As Buffon's oeuvre admirably demonstrates, the modern age cannot be described as that of a cleavage of times, of a gulf between planetary and social temporalities. On the contrary, it was an age when man's power over nature, his ability to divert its course, was contemplated with dread but even more with elation. The history of humanity stirred into its stories the climate, the frozen rivers of ancient Gaul, and the clearing of America, and would continue to do so far into the nineteenth century.

42 Buffon, *Histoire naturelle*, supplément, vol. 5, p. 240.

4

The Birth of Historical Climatology

The meteorological reports featured in papers at the end of the eighteenth century had a broad compass, covering all sorts of issues of the day: epidemics, the condition of harvests, frozen waterways putting factories out of work, dried-up rivers disrupting supplies to towns, and, of course, the frosts, hailstorms and droughts that heralded high prices.[1] The climate was perceived as intimately bound up with most human activities – agriculture and transport and, via them, trade as a whole, but also health by way of the neo-Hippocratic paradigm dominant in medicine. Much more so than for us, climate at that time amounted to a 'total social fact'. Detecting, measuring, and anticipating its fluctuations was of crucial importance.

In the 1770s, a genuinely historical approach to the climate emerged. The objective henceforth was to generate long-term temporal series, making it possible to identify a trend or demonstrate the stability of the climate. Observation logs were analysed, and sources consulted to extract meteorological information – major cold spells, droughts or storms – and the latter were tabulated by type of event, compared and criticized in order to assess their climatic significance. Scientists also studied the height of rivers, the changing character of vegetation, and the movement of glaciers.

1 From 1784 the *Journal de Paris* published a daily weather report, very popular with Parisians, derived from the Paris Observatory's observations (archives of the Paris Observatory, D6–39).

With their analysis of historical sources and recourse to 'proxies',[2] the scientists of the 1770s laid the bases for an investigation of the climatic past. In the nineteenth century, these methods would be taken up, criticized, and perfected by several generations of scientists (see Chapter 14). The science of historical climatology was thus not founded by the major figures – Emmanuel Le Roy Ladurie or Hubert Lamb – who embodied it in the twentieth century. It sprang from a long history mingling climate anxiety, the aspiration to anticipate the future, and a desire to write the interlinked history of peoples, living things and the Earth.

Meteorologists Tackle the Past

It sometimes happens that major innovations nestle in the most humdrum objects. This is exactly what occurred in meteorology in the late seventeenth century, with the introduction of the *observation log*. These simple notebooks, completed on a daily basis, represented a major change in the approach of scientists. The observer no longer selected specific phenomena but was obliged to log the condition of the atmosphere even when it presented nothing of note. The extraordinary gave way to the normal, curiosity to regularity.[3] The diligence, perseverance and elimination of personal judgement required for log-keeping, and the ritualization of measuring – at fixed hours – were set to lead to the revelation of the laws of nature. Scientific institutions encouraged this new personal discipline. In England, Robert Hooke, Boyle's former assistant who was in charge of the Royal Society's instruments, published a protocol stipulating correct observational methods in 1667. In France, regular meteorological observations were organized at the Paris Observatory, created the same year.[4]

2 That is, 'para-meteorological' phenomena, having some link to the prevalent weather and intelligible via historical sources.

3 On the links between scientific observation and the construction of subjectivities, see Vladimir Jankovic, *Reading the Skies: A Cultural History of English Weather, 1650–1820*, Chicago and London: University of Chicago Press, 2000; Fabien Locher, *Le Savant et la tempête.* Étudier *l'atmosphère et prévoir le temps au XIXe siècle*, Rennes: PUR, 2008; and Golinski, *British Weather and the Climate of Enlightenment.*

4 Robert Hooke, 'Method for Making a History of the Weather', in Sprat, *History of the Royal Society*, London: printed by T.R. for J. Martyn, 1667, pp. 173–8; Jean-Pierre Legrand and Maxime Le Goff, 'Louis Morin et les observations météorologiques sous Louis XIV', *Comptes rendus de l'Académie des sciences*, general series, vol. 4, no. 3, 1987, pp. 251–81; Richard C. Cornes, 'Early Meteorological Data from London

Quantitative meteorology gave a new meaning to the term 'climate'. It no longer referred exclusively to a space between two latitudes, but increasingly coincided with the ancient notion of zone, with the difference that binary oppositions (cold/hot, dry/wet) could now be objectified by means of instruments. During the eighteenth century, scientific literature began to use the word 'climate' in its contemporary sense, to refer to a certainty regularity of meteorological parameters. For example, the *Encyclopédie* contained two articles on 'climate' – the first in geography, which adopted the astronomical definition, and the second in medicine, which explained that '*climate*, in this sense, is exactly synonymous with *temperature*'[5] (others would include humidity or winds). Climate referred to a regularity concealed by erratic variations in the weather, and its definition by quantitative meteorology involved calculating *averages*, with the maximum possible number of observations and years of hindsight.

However, in the mid-1770s it appeared that this climate was itself changing in accordance with slow trends, identifiable over several decades. Thanks to the accumulation of meteorological logs over a century, meteorology could set out to tackle the past. The most prominent meteorologists of the time (Giuseppe Toaldo in Padua, Louis Cotte in Montmorency, Van Swinden in Holland, Théodore Mann in Brussels) made climate change an important issue for their discipline. And change became measurable, because meteorologists no longer confined themselves to calculating the overall average from their data but reconsidered that data diachronically. 'In a word,' wrote Louis Cotte, 'the meteorological observer must be the *historiographer* of nature.'[6]

This new attitude to logs stemmed from a scientific project highly fashionable in the 1770s: *astro-meteorology*. Its aim was to correlate the condition of the atmosphere with the respective positions of the Earth, the Moon, and the Sun. At stake was reducing meteorology to astronomy, in order to endow the first with the predictive power of the second. This project was careful to distinguish itself from astrology. There was no need to advance the hypothesis of some mysterious influence: it was the attraction of the celestial bodies and their heat that connected their courses

and Paris: Extending the North Atlantic Oscillation Series', doctoral thesis, University of East Anglia, 2010; Alfred Fierro, *Histoire de la météorologie*, Paris: Denoël, 1991, pp. 75–9.

5 *Encyclopédie*, article on 'Climate'.

6 Louis Cotte, *Traité de météorologie*, Paris: Imprimerie royale, 1774, p. 519.

to the Earth's atmosphere. On the tidal model, lunisolar configurations determined 'atmospheric tides'.

The astronomer Giuseppe Toaldo was the figurehead of this movement. In 1768, he inherited the meteorological registers from the Marquis de Poleni, a mathematician and astronomer at the Padua Observatory. They covered a period of exceptional length: 1725–68.[7] Toaldo then embarked on a meticulous analysis of these weather logs to uncover the relations between astronomical positions and the state of the weather. His goal was to identify the 'return of seasons', to pinpoint amid the numerical chaos stable temporal structures that brought the same atmospheric conditions in time. A seasoned calculator (and previously an author of trigonometric tables), Toaldo was well equipped to meet this challenge. The discovery of such periods would be extraordinarily beneficial: it would render meteorology wholly predictive, lending it immense practical interest.[8] Thus Toaldo obtained subsidies from the Venetian Senate, for his work would make it possible to anticipate bad harvests and so build up reserves.[9] He also reported collaborations with landowners in Frioul and financiers, helping them to forecast future grain prices.[10] According to an enthralled doctor, astro-meteorology might ultimately make it possible 'to predict the return of epidemics, just as one predicts that of a comet'.[11] Toaldo's writings were distributed, translated, and had a Europe-wide impact.[12] And when, in the 1770s, the first major observation campaigns were launched, orchestrated by scientific societies (the Societas meteorologica palatina in Mannheim, the Royal Societies of Medicine and Agriculture in Paris), one of their objectives was precisely to record phenomena over the long term so as to solve the mystery of meteorological recurrence.[13]

7 By comparison, Cotte's observations at Montmorency only ran for thirteen years, from 1769–82.

8 The words used ('conjectures', 'rules of foresight'), and the reference to Bernoulli's theorem by Giuseppe Toaldo, attest to the desire to demarcate himself from sellers of almanacks. In 1805, drawing on Toaldo, Cotte published weather forecasts extending to the end of the nineteenth century. See Louis Cotte, *Mémoire sur la période lunaire de dix-neuf ans*, Paris: Hazard, 1805.

9 Giuseppe Toaldo, *Essai météorologique sur la véritable influence des astres, des saisons et changemens de tems*, Chambéry: Gorrin, 1784, p. 189.

10 Ibid., preface, pp. xvii, 220–222.

11 Ibid., preface, p. xii.

12 *Observations sur la physique*, 1777, vol. 10, pp. 243–79, 333–67; *Journal des savants*, December 1786, p. 815. The treatise was translated into Italian, Spanish and German.

13 Louis Cotte, *Mémoires sur la météorologie*, Paris: Imprimerie royale, 1788, vol. 1, p. x.

But what matters to us is the following: it was within the astro-meteorological paradigm that the first quantitative characterization of a changing climate emerged. To discover cycles, Toaldo used an instrument destined for a great future in the study of climate: deviation from the mean. In 1770, after comparing annual thermometric averages in Padua with the average calculated over forty years, he noted, in passing, that the former had fallen over the last twenty years.[14] Louis Cotte, who wrote a laudatory review of this work, also lent credence to the idea of a recent 'cooling of the seasons'.[15] In 1774, Toaldo published a new table showing significant cooling (5° C) in Padua between 1725 and 1774.[16] For the first time, meteorologists were producing a diachronic view of a local climate over several decades.

The Pitfalls of Historical Thermometry

Toaldo's work was based on handling a great deal of data. He acknowledged that 'some lack of attention may have slipped into the observations, the thermometers, the situation of the place, the hour . . . but, nobody, I believe, would make so bold as to deny everything.'[17] He sought here to parry criticisms arising from the discrepancy between his 'statistical' methods and the experimental culture of precision, then on the rise in the scientific world.[18] Could thermometers – old thermometers, indeed – really be trusted to measure climate change?

In Paris, a particular place crystallized this doubt: the deep cellars of the Observatory. Since the end of the seventeenth century, the scientific consensus had been that their temperature was constant at 10° Réaumur. They even served as a fixed point for calibrating thermometers.[19] Yet, in 1774, using old instruments that had been calibrated on the spot, two astronomers

14 Giuseppe Toaldo, *Della vera influenza degli astri. Saggio meteorologico*, Padua: Manfre, 1770, p. 146.

15 *Journal des savants*, October 1771, p. 103.

16 Giuseppe Toaldo, *Essai de météorologie appliquée à l'agriculture*, Montpellier: Martel, 1774, p. 58 and *Meteorologia applicate all agricultura*, Venice: Storti, 1775, p. 40.

17 Toaldo, *Essai météorologique*, p. 174.

18 Norton Wise, ed., *The Values of Precision*, Princeton: Princeton University Press, 1995 and Christian Licoppe, *La Formation de la pratique scientifique: le discours de l'expérience en France et en Angleterre, 1630–1820*, Paris: La Découverte, 1996.

19 Jean Gaussen, *Dissertation sur la thermomètre de Réaumur*, Béziers: Fuzier, 1789, p. 227.

noticed that the cellars had cooled by 1° Réaumur in forty years.[20] These results were striking, as they confirmed Buffon's theory of the cooling of the globe. Toaldo also hastened to comment on them, for he regarded them as corroborating the drop in temperatures measured in Padua.[21]

The question was revived by the heavy winter of 1775–76, which itself followed a series of bad harvests that had sparked serious food riots: the famous Flour War of 1775.[22] During this freezing winter, people became captivated by meteorology. The astronomer Charles Messier, who kept a meteorological diary, complained of being bothered every morning by 'a crowd of Parisians wishing to know the temperature'.[23] At the Académie des sciences, a committee composed of major experimenters (including Lavoisier) was tasked with comparing the cold of 1775 with that of 1709, another terrible winter. According to the academician Baumé, it was 'society people' who earnestly desired that they should be compared.[24] In addition to curiosity about the extreme phenomenon, the question raised was the frequency of 'heavy winters' and hence the evolution of the climate. The exceptional cold prompted concerns about the very stability of the natural order. It might be confirmation of the thermal death of the globe or, indeed, the sign of a 'disruption of the seasons'. For one journalist, the European climate had changed since the Lisbon Earthquake, because, he explained, 'everything is connected in the great physical machine of the globe. The slightest alteration in one of its parts is bound to produce consequential ones in all the others.'[25]

But these debates would, above all, sow the seeds of doubt about the reliability of old thermometers. The line of argument from the cold detected in Padua or the cellars of the Observatory seemed too hasty, even for Buffon's defenders. According to Louis Cotte, the deviation measured in the Parisian cellars was more likely to be explained by an alteration in the

20 *Observations sur la physique*, 1774, vol. 4, p. 480 and *Histoire de l'Académie royale des sciences*, year 1774, Paris: Imprimerie royale 1778, p. 688.

21 *Observations sur la physique*, 1778, vol. 13, p. 456.

22 Le Roy Ladurie, *Histoire humaine et comparée du climat*, vol. 2, pp. 28–103. The winter of 1772 was disastrous and, three years later, the harvests were again inadequate.

23 Charles Messier, 'Mémoire sur le froid extraordinaire qu'on a ressenti . . . au commencement de cette année 1776', *Histoire de l'Académie royale des sciences*, year 1776, 1779, p. 63.

24 Baumé, *Opuscules chimiques*, Paris: Agasse, Year VI, p. 214.

25 Simon Linguet, 'Du dérangement dans l'ordre des saisons', *Annales politiques, civiles et littéraires*, vol. 1, 1778, pp. 477–89.

liquid of the thermometers.[26] According to Charles Messier, it was above all because, prior to the 1770s, people descended into the cellar with torches that disrupted the instruments.[27] The committee of the Académie charged with the cold of 1775–76 discovered to its amazement that thermometers were generally poorly calibrated.[28] Thirty-eight thermometers 'deemed excellent' all measured different temperatures. 'It must be agreed', Baumé confessed, 'that scientists . . . led astray for more than fifty years . . . have been caught off guard.'[29]

The rise in the last third of the eighteenth century of the culture of precision led to doubt being cast on old thermometric measurements. How, then, without the aid of instruments, was one to work towards the historical reconstruction of the climate?

The Sources of Historical Climatology

The first solution, and the most obvious one, was to extract from historical chronicles the meteorological information they contained: 'heavy winters', 'extraordinary colds', droughts, floods, and so on. This is what Giuseppe Toaldo and Charles Messier did in 1776 following the thermometric controversies.[30] Taking Toaldo as his guide, the Viennese Jesuit

26 Cotte, *Mémoires de météorologie*, vol. 2, pp. 482–5.

27 Messier, 'Mémoire sur le froid', p. 42.

28 Bézout, Lavoisier and Vandermonde, 'Expériences faites par ordre de l'Académie, sur le froid de l'année 1776', *Mémoires de mathématique et de physique tirés des registres de l'Académie royale des sciences*, year 1777, 1780. Cf. Jean-François Gauvin, 'The Instrument that Never Was: Inventing, Manufacturing, and Branding Réaumur's Thermometer during the Enlightenment', *Annals of Science*, vol. 69, no. 4, pp. 515–49.

29 Baumé, *Opuscules chimiques*, p. 214. If the academicians devoted protracted experiments, and not less than four memoirs, to measuring the cold of 1775, it was also because the latter revealed the instability of thermometers on which numerous practices, scientific (from chemistry to positional astronomy) and commercial (yardsticks were established at 12° Réaumur), were dependent.

30 Giuseppe Toaldo, 'Discorso sopra l'anno 1776', *Giornali astro-meteorologici dall' anno 1773 all anno 1798*, Venice: Francesco Andreola, 1802, vol. 1, pp. 138–47 and *Essai météorologique*, pp. 244–53. We may note that from the mid-eighteenth century, doctors published meteorological chronicles whose aim was not to study the development of climate, but to make an inventory of the conjunctions between epidemics and meteorology. Cf. Thomas Short, *A General Chronological History of the Air, Weather, Seasons, Meteors &c. in Sundry Places and Different Times*, London: Longman, 1749; John Rutty, *Chronological History of Weather and Seasons and the Prevailing Diseases in Dublin*, London: Robinson, 1770.

and astronomer Anton Pilgram published an enormous tome on the history of the climate in 1788. He employed no less than 108 sources, whose meteorological information he presented in tabular form: cold and mild winters, wet and dry years, violent winds, and the good and bad years for wine.[31] Finally, in 1792 Abbé Théodore Mann of the Brussels Academy published an important *Mémoire sur les grandes gelées et leurs effets*.[32] Generally aiming to refute Buffon's theory, he divided heavy winters into seven categories of intensity and drew two conclusions. Firstly, harsh winters were more common in remoter periods, suggesting that the Earth was growing warmer. Secondly, historical study did not detect any periodic return of heavy winters – something that torpedoed astro-meteorological theories.[33]

But the problem that immediately arose once again was the reliability of past accounts. According to the Dutch physician Van Swinden, one should be wary of the exaggeration of chroniclers. Conversely, changing sensitivities had to be taken into account: what counted as a mild winter for an author from antiquity might well be glacial in eighteenth-century terms. Historical climatology needed to update its methods: 'criticism is no less necessary in physics than in literature.'[34] Charles Messier stressed the difficulty of the undertaking. Having drawn up a table of instances of the Seine freezing over, as listed since 1392 in the *Chronicle of Saint-Denis*, he immediately queried the meaning of his work. Comparing these data with thermometric measurements for the years 1760–70, he noted that there was no straightforward relationship between air temperature and the river freezing. Further determinants included the water level and flow, the air's humidity, and other, unknown factors.[35]

Despite these uncertainties, rivers became climatic proxies, at least as regards precipitations. This was because data was available: water levels

31 Anton Pilgram, *Untersuchungen über das Wahrscheinliche der Wetterkunde durch vieljährige Beobachtungen*, Vienna: Joseph Edlen, 1788.

32 Théodore Mann, *Mémoire sur les grandes gelées et leurs effets, où l'on essaie de déterminer ce qu'il faut croire de leurs retours périodiques et de la gradation en plus ou moins froid de notre globe*, Ghent: Goesin, 1792. This dissertation has been republished with an introduction by Muriel Collart and a preface by Emmanuel Le Roy Ladurie, Paris: Hermann, 2012.

33 Ibid., p. 112.

34 Van Swinden, 'Lettre sur les grands hivers adressée au citoyen Cotte', *Journal de physique et de chimie*, 1800, vol. 1, pp. 279–81.

35 Messier, 'Mémoire sur le froid', pp. 63–79.

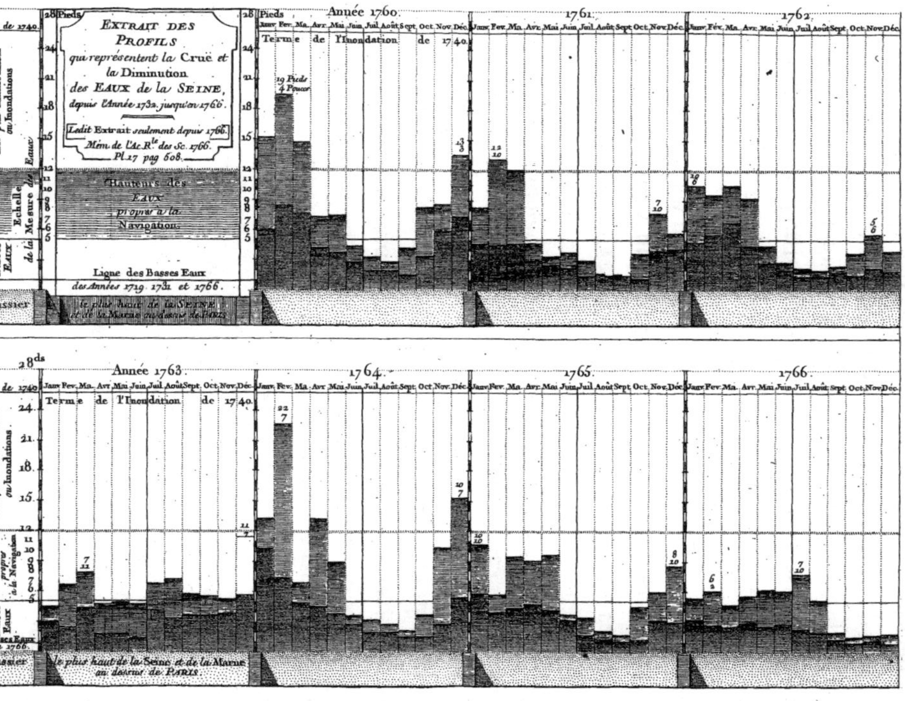

In 1776 Paris commissioned a historical study of the Seine from Philippe Buache, the king's chief geographer. Buache, 'Exposé des divers objets de la géographie physique, concernant les bassins terrestres des fleuves et rivières qui arrosent la France, et en particulier celui de la Seine', *Mémoires de l'Académie des sciences*, 1767, p. 508.

Bibliothèque nationale de France, Paris

were so important for supplying towns that they had long been subject to meticulous monitoring. Thus, from 1732 the level of the Seine was measured every day by the Paris City Office. It was therefore on the basis of a solid corpus that 'histories of floods' were written,[36] or that the meteorologist Louis Cotte could assert that the drought of 1800 in Paris was probably the most severe of the eighteenth century.[37]

The vegetable kingdom furnished a second proxy. In the late eighteenth century, naturalists took a close interest in the relations between plants and climate. With great disappointment, they accepted that they would not be able to acclimatize rice or palm trees in Northern Europe, as Carl von Linnaeus had hoped in the middle of the century. On the contrary: the experience of greenhouses demonstrated the great sensitivity of vegetal phenomena to differences in climate. Likewise, the first studies in botanical geography were focused on the location of vegetation. One of the pioneers in the field, Father Jean-Louis Giraud Soulavie, defined 'climates' – in the sense of region – in accordance with the plants predominant at certain altitudes.[38] Meteorologists also sought to make themselves useful to farmers: Cotte, for example, measured the total summer temperatures required for grapes to ripen.[39]

The important point is that all this knowledge crossing meteorology with botany transformed any information about past vegetation into so many signs of climate change. For example, if in the sixteenth century grape harvests had occurred a month earlier, this could indicate that

36 Antoine Deparcieux, 'Mémoire sur les inondations de la Seine', *Mémoires de l'Académie des sciences*, year 1764, Paris: Imprimerie royale, 1767, pp. 457–87.

37 Louis Cotte, 'Notice des grands hivers . . . et des grandes inondations de la Seine à Paris', *Journal de physique*, vol. 48, Nivôse Year VII, 1799, pp. 270–80 and 'Notes sur la chaleur et la sécheresse extraordinaires de l'été de l'an VIII (1800)', *Journal de physique*, vol. 51, Messidor Year VIII, p. 216.

38 For example, in the case of the mountains of the Vivarais, Soulavie identified six climates (in the sense of region), defined by their dominant species: orange tree, olive tree, grape vine, sweet chestnut, conifer and alpine grass. Cf. Jean-Louis Giraud Soulavie, *Histoire naturelle de la France méridionale. Les végétaux*, Paris: Quillau, 1783, vol. 2, p. 285. Marie-Noëlle Bourguet, 'Landscape with Numbers: Natural History, Travel and Instruments in the Late Eighteenth and Early Nineteenth Centuries', Bourguet, C. Licoppe and O. Sibum, eds, *Instruments, Travel and Science: Itineraries of Precision from the Seventeenth to the Twentieth Century*, London: Routledge, 2002, pp. 97–126.

39 Cotte, *Traité de météorologie*, pp. 420–2, 451–7. G. Pueyo, 'La météorologie à la Société royale d'agriculture au cours de la seconde moitié du XVIIIe siècle', *Comptes rendus des séances de l'Académie d'agriculture de France*, 1975, pp. 217–26.

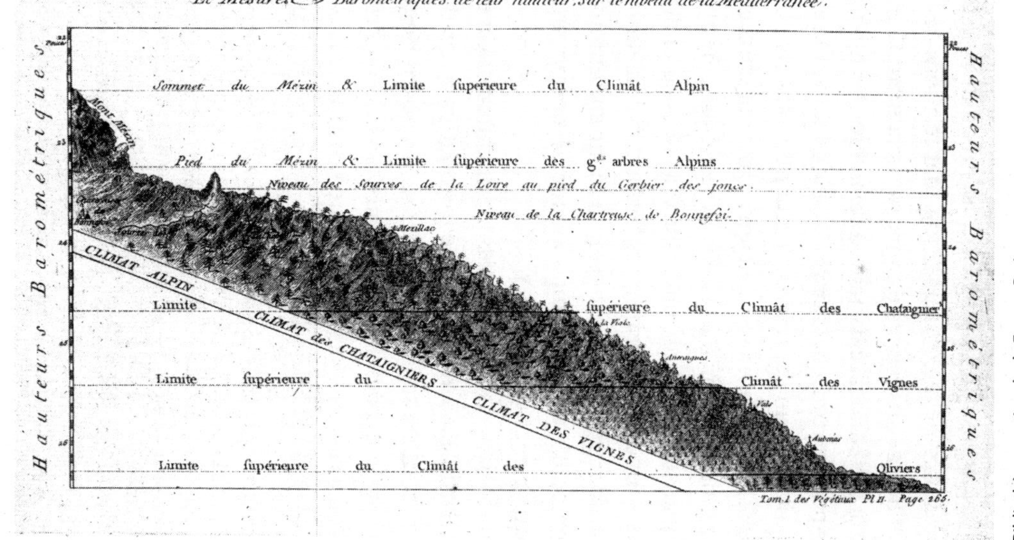

The botanical definition of climates. 'Coupe verticale des montagnes vivaroises avec les limites respectives et les mesures barométriques de leur hauteur'. Jean-Louis Giraud Soulavie, *Histoire naturelle de la France méridionale*, vol. 2, *Les Végétaux*, Paris, 1783, p. 265.

European summers had since become colder.[40] One phenomenon was particularly intriguing: the migration of certain plants over the course of time. In Switzerland, botanists noted with disquiet that, in the past, trees had flourished above the current tree line.[41] Was this evidence of encroaching cold? Soulavie was surprised to find himself paying a cens in wine on certain of his properties where viticulture was impossible. Intrigued, he consulted an expert on feudalism: it seems that in the fourteenth century vines did grow there, suggesting that the climate had become much colder.[42] Seigneurial rights payable in certain crops, but no longer corresponding to the local territory, attested to the historicity of the climate. We find the same questions in England in connection with the existence of viticulture in the twelfth century.[43]

Third and last climatic proxy: glaciers. These enjoyed great cultural visibility at the time. Having set out on their Grand Tour, on the way to Italy young aristocrats came upon the Alpine range, with its vertiginous summits, immense panoramas and impressive glaciers. The glaciers perfectly embodied the sublime theorized by Edmund Burke: nature as grandiose, implacable and terrifying. Countless paintings and engravings brought the experience vicariously to life.[44]

And yet in the late eighteenth century, in conjunction with Buffon's theory of the Earth, glaciers symbolized the fragility of every being faced with the Earth's inexorable cooling. Marc-Théodore Bourrit, a Swiss engraver and alpinist, who, more than anyone else, contributed to the European enthusiasm for glaciers, explained that the *increase in ice* was the question of greatest interest to visitors.[45] The encroaching ice was a

40 'Observations sur la chaleur des climats par M***, gentilhomme du Vivarais', *Observations sur la physique*, 1774, vol. 3, pp. 245–51.

41 Reynier, 'Mémoire sur l'abaissement de la région boisée', *Mémoires d'agriculture, d'économie rurale et domestique*, 1790, p. 63.

42 Soulavie, *Histoire naturelle de la France méridionale*, p. 241.

43 Samuel Pegge, 'The Question Considered Whether England Formerly Produced Wine from Grapes', *Archaelogia*, 1775, vol. 3, pp. 53–95.

44 Marjorie Hope Nicholson, *Mountain Gloom and Mountain Glory: The Development of the Aesthetics of the Infinite*, Ithaca: Cornell University Press, 1959. On the scientific 'discovery' of mountains in the eighteenth century, see Numa Broc, *Les Montagnes au siècle des Lumières*, Paris: CTHS, 1969 and Peter H. Hansen, *The Summits of Modern Man: Mountaineering after the Enlightenment*, Cambridge, MA: Harvard University Press, 2013.

45 Marc-Théodore Bourrit, *Guide du voyageur aux glaciers de Chamouni*, Geneva: Abraham, 1828, vol. 1, p. 187 and *Nouvelle Description des glacière et glaciers de Savoye*, Geneva; Paul Barde, 1785, vol. 2, p. 196. Let us note that the encroaching ice was already

mandatory chapter in all Alpine narratives. Hiking the Chamonix valley and observing it first-hand amounted to witnessing prodromes of the globe's thermal death. According to Buffon, 'the expansion of these lands of ice is already, and will subsequently be, the most tangible proof of the consecutive cooling of the Earth.'[46] While the reports, paintings and engravings of glaciers became objects of cultural consumption, the theory of global cooling gave them an eschatological significance. They were the *memento mori* of nature in its entirety.

Alongside – or against – this end-of-the-world aesthetic, several learned Swiss figures laid the bases of glaciology. Their testing ground was the Grindelwald valley, not far from Bern. Drawing on multiple sources (oral testimony, charters, maps, field observations), they succeeded in reconstructing the valley's climatic history with remarkable precision. There had been a phase of rapid glacier growth in the late seventeenth century: at this time, wrote one, 'the inhabitants were alerted by an extraordinary development . . . nature departed from its normal course.'[47] There followed a temporary retreat until 1750, and then a new expansion. The scholars' approach was purely historical: glaciers were proxies for complex climatic variability, conforming to no pre-established theory of the Earth. The jagged glacial chronology belied the thesis of progressive cooling. Added to this was the fact that the link between climate and glaciers was anything but clear. Their expansion might only be apparent, merely an effect of their mass making them drain slowly 'like softened beeswax.'[48] At the Académie des sciences in Paris, where Buffon's theory had its enemies, some subscribed to this thesis: the advance of glaciers was in fact attributable to their *melting* and the resulting slippage.[49]

---

well-noted – and suffered – by mountain communities from the mid-seventeenth century. Cf. Raoul Blanchard, 'La crue glaciaire dans les Alpes de Savoie au XVIIe siècle', *Recueil des travaux de l'institut de géographie alpine*, vol. 1, no. 4, 1913, pp. 443–54; Le Roy Ladurie, *Histoire humaine et comparée du climat*, vol. 1, 'Canicules et glaciers, XIIIe siècle–XVIIIe siècle', pp. 409–529.

46 Buffon, *Les Époques de la nature*, vol. 1, pp. 143–4.

47 Bernhard Friedrich Kuhn, 'Versuch einer Beschreibung des Grindelwald Thales', *Magazin für die Naturkunde Helvetiens*, Zurich, 1787, pp. 117–36. See also Gottlieb Sigmund Grüner, *Die Eisgebirge des Schweizerlandes*, Berne: Wagner, 1760; French translation, *Histoire naturelle des glaciers de Suisse*, Paris: Panckoucke, 1770, pp. 329–34.

48 César Bordier, *Voyage pittoresque aux glaciers de Savoye fait en 1772*, Geneva: La Caille, 1773, p. 225.

49 'Sur le froid de 1776', *Histoire de l'Académie des sciences*, year 1776, Paris: Imprimerie royale, 1779, p. 13; Nicolas Desmarest, 'Précis d'un mémoire sur le mouvement

In 2006, during the collective elaboration of the fourth report of the IPCC,[50] the climatologist Stefan Brönnimann proposed to include the works of scientists of the early modern era, notably those of Abbé Mann. The IPCC rejected this inclusion on the grounds that it did not 'really involve modern science'.[51] In one sense, our investigation vindicates the IPCC. It was not greenhouse gas emissions that prompted seventeenth- and eighteenth-century scientists to concern themselves with climate change, but a thousand other things – colonizing North America, unlocking the secret of meteorological cycles, forecasting the thermal future of the Earth, and understanding the impact of human action on the vegetable world, the water cycle and, through it, the climate. However, this division seems far too absolute. Contemporary thinking and knowledge of climate change inherit a long history. The emergence of historical climatology is rooted in an intellectual moment when human action and the destiny of planet Earth were intimately interwoven.

progressif des glaces dans les glacières et sur le phénomènes qui dépendent de ce mouvement successif', *Observations sur la physique*, vol. 13, 1779, p. 383.

50 Intergovernmental Panel on Climate Change.

51 pds.lib.harvard.edu/ (accessed 21 February 2014).

5

An Arsenal in the Indian Ocean

Until the end of the eighteenth century, the North American colonial experience structured reflection on climate change: with their axes and ploughshares, the colonists were making the climate milder. The prevalent idea was a virtuous circle linking culture, climate, and civilizing process. A grand narrative at once historical, natural, and moral presented North America as a wooded, cold and wet space, similar to Northern Europe prior to the Roman conquest. In this regard, travelling to America was also travelling in time, back to the origins of civilization. This climatic grand narrative was supported by advances in plant physiology. Research on the 'transpiration of plants' initiated by John Woodward (Chapter 2) was continued by Stephen Hales in England and by the agronomist and forestry expert Duhamel du Monceau in France. They reinforced the idea of vegetation's humidifying and cooling effect.[1]

As a result, in the mid-eighteenth-century, Europe's forests were still regarded as a distinctly negative climatic factor. In the 1760s, the peasants

1 Stephen Hales, *Vegetable Statistics*, London: Innys, 1727. This book, which was fundamental for the study of vegetal physiology and the phenomenon of evapotranspiration, was translated by Buffon in 1735. Cf. Grove, *Green Imperialism*, pp. 153–67. Duhamel du Monceau, *La Physique des arbres où il est traité de l'anatomie des plantes et de l'économie végétale*, Paris: Guérin et Delatour, 1758. Duhamel du Monceau and Buffon recommended cutting down copses situated near sensitive cultivation like vines or fruit trees. Cf. 'Observations des différents effets que produisent sur les végétaux les grandes gelées d'hiver et les petites gelées de printemps', Œuvres *complètes de Buffon*, vol. 2, Paris: Pillot, 1838, p. 365. See also Duhamel du Monceau, *De l'exploitation des bois*, Paris: Guérin, 1764, vol. 1, p. 99.

of the Alpine foothills blamed walnut trees for 'attracting lightning and entrenching humidity'.[2] Even pleas for forest conservation barely employed the climatic argument or, indeed, warming was presented as scant consolation for the loss of forests and wood resources.[3] In their 'medical topographies', doctors too looked on forests with suspicion, because they allegedly blocked the circulation of air.[4] As late as 1774, the article on 'Rain' in the *Encyclopédie* reiterated this pejorative view. In it we read that the 'immense forests of Sweden cause such abundant rains that they . . . destroy fertility'. The situation, added the author, was the same in the Caribbean Islands prior to the arrival of the Spanish. At a distance of nearly three centuries, we find Columbus's climatic argument in favour of colonization, now applied to Scandinavia.[5]

A Nature for War

The reversal of climate discourse on forests occurred principally in France in the late eighteenth century. It stemmed from several causes, which are explored in this chapter and the next. The first was the novel importance of the tropical island in naturalist culture. Thinking about the climate changed direction, for it switched its main location: from the wintry weather of North America, scientific speculation now turned towards the tropical spaces where drought threatened. The 1760s were crucial. We know that English and French exploration of the Pacific Ocean made tropical islands the matrix of moral, social, and sexual utopias, but also of *environmental* utopias. Just as the 'savage' served as

2 AN F20 154, Thomas Riboud de Bourg, 'Mémoire sur le département de l'Ain relativement à la culture et sur quelques espèces qui y deviennent rare', Paris, Year VIII (February 1800).

3 Robert Hinckmann, 'La pratique des enclos, adoptée en Angleterre, est-elle avantageuse aux défrichements?', *Mémoires sur les questions proposées par l'Académie des sciences et belles lettres de Bruxelles*, Brussels: Boubers, 1774, p. 82.

4 Boucher, 'Description du climat de la ville de Lille en Flandres', *Journal de médicine, chirurgie, pharmacie*, 1757, vol. 7, p. 234. Even in hot climates, the cooling – hence beneficial – effect of forests seemed to be counterbalanced by the obstacle they created to the circulation of air. Cf. *Lettres édifiantes et curieuses écrites des missions étrangères par quelques missionnaires de la compagnie de Jésus*, 18th collection, Paris: Guérin 1758, p. 47; MacCartney, *Voyage dans l'intérieur de la Chine*, vol. 1, Paris: Buisson, 1798, p. 207.

5 *Encyclopédie, ou dictionnaire universel raisonné des connaissances humaines*, Yverdon, vol. 34, p. 173.

a model for a humanity uncontaminated by laws and education, so the island became the key site for conceiving a state of nature prior to the action of civilized man.[6] In scientific and literary culture, and beyond, the idea of a paradisiacal nature preserved over there, on the other side of the Earth, took root.

In *Green Imperialism*, the historian Richard Grove defends the idea of the colonial and insular genesis of environmentalism. Eighteenth-century naturalists working in the colonies were probably the first people to have envisaged the large-scale, deleterious impact on nature of human action – thanks to the unique viewpoint on the world afforded by insular spaces. In a confined place, isolated from the outside, the effect of deforestation on the environment would have been immediately visible. Ecological sensibility is said to have been born here, from the encounter between this full-scale experience and a naturalistic culture blended with Rousseauism, the myth of the noble savage, and Physiocracy.[7]

One figure occupies a central place here: Pierre Poivre, a French missionary and naturalist who was also intendant of the Île de France (today's Mauritius) between 1767 and 1772. He it was who issued the very first climate warning in history. The event was important enough for Grove to elevate it into practically the birth certificate of environmentalism. On Mauritius, Poivre is said to have invented a new, holistic view of nature, connecting the virtual disappearance of rainfall from the island with deforestation: a decisive step relative to older conceptions of forest conservation, which regarded woodlands as mere stocks of timber to be maintained the better to exploit them. This theory spread to English and Dutch colonies, thanks to botanical networks, and later to Europe and North America.

Grove thus makes Poivre a hero in the ecological history of the West, adorning him with a thousand virtues: opponent of slavery, Rousseauist, Sinophile and *sinisant*, a genuine connoisseur of Chinese agriculture or the forestry policies of Mughal India. Equipped with these credentials, Poivre, combining European plant physiology (Woodward, Hales, Duhamel) with Eastern knowledge of forestry, developed the first environmental policy in history: protecting the climate by defending forests. In the intellectual

6 Michèle Duchet, *Anthropologie et Histoire au siècle des Lumières*, Paris: Flammarion, 1977, p. 13.

7 Grove, *Green Imperialism* and Gregory Quenet, 'Protéger le jardin d'Éden', in Richard Green, *Les Îles du Paradis. L'invention de l'écologie aux colonies, 1660–1854*, Paris: La Découverte, 2013, pp. 77–120.

context of the 1990s, marked by the rise of global and environmental history, Grove's appreciation of this thesis is understandable.

Yet it is problematic. From a meteorological point of view, the role of trees in encouraging rainfall is lesser in insular climates than in continental ones, with most humidity deriving from the evaporation of the oceans. The island is therefore not *naturally* a 'laboratory' for climatic reflection, but rather the opposite. Above all, a more precise study of Pierre Poivre's activity leads to a very different interpretation of his climate concerns.[8] According to the orders he received from Versailles, Poivre was to implement a radical transformation of the Île de France, converting it from a plantation economy into a powerful arsenal intended to expel the British from India. As we will see, it was to justify this abrupt conversion that Poivre resorted to climate alarmism.

Let us go over the file. Emerging from the Seven Years' War (1756–63) won by Britain, the Compagnie des Indes orientales that administered the France and Bourbon islands (Mauritius and Réunion) was bankrupt. The Crown decided to take back control of these territories. The choice of Poivre as intendant – an eminently strategic post in the context of Franco-British rivalry in the Indian Ocean – made sense: as an agent of the Compagnie des Indes, he was already familiar with the colony's problems. Most importantly, he was close to the Duc de Choiseul – the de facto head of government under Louis XV – as well as his cousin Choiseul-Praslin, the navy minister.

On his arrival on the Île de France in July 1766, Poivre delivered a thundering speech articulating the king's grievances against the colonists: corruption, cohabitation with slaves, cultivation of sugar and coffee at the expense of cereals, absentee landlords, and so on. Through the fault of the colonists the island was, he declared, on the verge of a moral and economic catastrophe. In this harsh indictment, deforestation loomed large. The settlers were 'greedy and ignorant men [who] have ravaged the island by destroying its woods, eager to make a quick fortune at the colony's expense'. Result, they 'have left their successors nothing but arid land, abandoned by the rains. Nature has done everything for the Île de France: men have wrecked everything on it. The magnificent forests that used to cover the ground once unsettled the passing clouds with their motion and caused them to dissolve into fertile rain.'[9] In 1769, an 'eco-

8 Most of the archives concerning Pierre Poivre are accessible online thanks to the valuable transcription work of Jean-Paul Morel: pierre-poivre.fr.

9 *Discours prononcé par M. Poivre, Commissaire du roi*, London, 1767. We should note the consistency with which Poivre referred to the problem of climate change. Thus, he spoke of it to Lyon's Académie royale des sciences in 1763 and the issue was also

nomic regulation' decreed lofty measures of forest conservation: the need for prior authorization of land clearance accompanied bans on building wooden houses; carrying 'brands into fields, even on the pretext of lighting a pipe'; and clearing hillsides, riverbanks and paths, as well as the island's coastline. Finally, at least one-quarter of leased land had to be left wooded, and reforested as and when it became necessary. The preamble justifying these measures mentioned the meteorological function of the forest, which 'protects crops against the violence of the winds, the heat of the sun and droughts'.[10]

How is this attention to the climate and forestry, unusual for its day, to be interpreted? Does it really involve warning of an imminent catastrophe? In fact, paradoxically, while Poivre invoked the climate, no shortage of wood, then or later, justified the measures. In a memorandum sent in 1767 to Minister Praslin, he indicated that of the island's 400,000 acres of cultivable land, 350,000 were still covered by forest.[11] 'Were we to put aside only 100,000 acres of wood, that is, one-quarter of the total surface area, that reserve would be sufficient.'[12] Plainly put, Poivre painted an apocalyptic picture of deforestation for the colonists, whereas, in his correspondence with the minister, he reckoned it possible to fell two-thirds of the remaining trees! So, it was not because the Île de France was on the brink of catastrophe that he issued his alert, but quite the reverse: despite abundant forests, it was necessary to check the land hunger of colonists wishing to extend their plantations.

For – and this was the crux of the matter – the government had tasked Poivre with a geostrategic mission: to transform the island into 'a formidable arsenal',[13] intended to exact revenge on the British and strip them of control in India.[14] Poivre even dangled before the minister the possibility of conquering China from the Île de France![15]

broached in the official instructions given him by Louis XV, which he probably wrote himself. Cf. A.N. Col C/4/17 f° 3–22. 'If the woods do not grow back, the rains will be less frequent and the overly exposed ground will be burnt by the sun.'

10 'Ordonnance de police n° 183. Règlement concernant la préservation des forêts. A l'Isle de France, le 15 novembre 1769', in Delaleu, *Codes des Iles de France et de Bourbon*, Port-Louis: Tristan Mallac, 1826, pp. 222–4.

11 30 November 1767, Poivre to Minister Praslin, Brest, Service historique de la Défense, département Marine. Ms. 89, n° 57.

12 Archives nationales, A.N. Col C/4/18, f° 381.

13 'Récapitulation du compte rendu par le Sieur Poivre Intendant de la Marine, de son administration des Isles de France et de Bourbon', A.N. Col C/4/30 f°231.

14 Poivre to Minister Praslin, 30 November 1767, A.N. Col C/4/18, f° 216.

15 Ibid., 18 February 1769, A.N. Col C/4/25, f° 28.

Twenty years earlier, in the colony's early stages, following a drought and shortage of grain, Mahé de La Bourdonnais, the first governor of the island, had had the greatest trouble arming a squadron against the British. Poivre admired Le Bourdonnais, victor in the naval battle of Negapatam (1746), a veritable hero and martyr for eighteenth-century French imperialists. In the very year of Negapatam, it was with Le Bourdonnais, returning from a victorious expedition to Madras, that Poivre had first discovered the Île de France.[16]

He wanted to accomplish what his model had been unable to achieve. The island was the French monarchy's great asset in the Indian Ocean: it threatened the British in India and also, thanks to the acclimatization of cloves, was set to wreck the Dutch monopoly on the spice trade. All of Poivre's activity reflected this desire for power. In 1772, when he departed from the island, he boasted about leaving a veritable arsenal behind: a tannery, forges and lime kilns (major consumers of wood), powder, granaries, a bakery and so on – enough to fully equip the 'strongest squadrons'.[17]

What is the relationship between the geostrategic role assigned to the Île de France and its climate? Firstly, by invoking a disastrous desiccation, Poivre sought to contain the colonists' appetite for land and divert them from sugar and coffee cultivation, which consumed space and wood. It also enabled him to justify the sizeable forestry reserves intended for the royal navy. A temperate climate, moreover, sheltered from storms, was a precondition for growing cloves as well as wheat. Finally, the temperate character of the climate would favour colonization by the French. In 1767, there were a mere 2,342 whites to 18,100 slaves. Over and above foodstuffs, powder and wood, the Île de France was to produce peasant-patriots ready to defend their land and, if necessary, conquer others in the king's name. The most striking thing about Poivre's eloquent speech to the settlers in July 1767 is its linkage of climate, critique of slavery, settlement colony and military force: 'The Île de France, *situated under a temperate sky* . . . was only to be cultivated by free hands. Its colonists were to be citizens drawn from the class of labourers in the metropolis; they would have been its formidable defenders.'[18] Poivre's climatic anxiety is the mirror of his racial anxiety and will to conquest.

16 La Bourdonnais, *Mémoires historiques de B.F. Mahé de la Bourdonnais*, Paris: Pélicier, 1827, p. 64.

17 Archives nationales, A.N. Col C/4/18, f° 398; Col C/4/30, f° 242.

18 Poivre, *Discours prononcé by M. Poivre.*

But what of the origins of Poivre's thinking about climate? Grove presents it as a radical innovation. Poivre was supposedly original in combining a range of influences: the island utopia, Physiocratic thought, Hales's plant physiology, and even various enigmatic 'Indian, Zoroastrian and Chinese systems of knowledge',[19] encountered during his missionary youth in China.

While Poivre was original in legislating on the forests of Mauritius for the good of the climate, the justification was scarcely original in itself. In the late seventeenth century, for example, John Evelyn interpreted the marvel of the sacred tree in a catastrophist sense (Chapter 2). In the following century, the prodigy was sufficiently familiar for the Abbé Prévost, the celebrated author of *Manon Lescaut*, to allude to it in a novel of 1744, where he wrote: 'as the forests are razed, the clouds and, consequently, the rains become scarcer and less heavy.'[20] Poivre also knew the famous story of the felling of clove trees on the Maluku Islands – ordered in 1652 by the Dutch, intent on preserving their monopoly – and its disastrous climatic impact.[21]

Moreover, in 1763, at the very moment when Poivre was speaking for the first time of climate change, an important text by the Jesuit Louis-Bertrand Castel, 'De l'action des hommes sur la nature', was republished.[22] Castel proposed a view of the Earth where everything, absolutely everything, and especially climate disasters, was conditioned by human action. For example, by dint 'of softening the Earth's external surface (with ploughing), we give rise . . . to volcanos, earthquakes, extraordinary thunderstorms, disturbance of the seasons.' The global water cycle linked all peoples in a zero-sum game: 'A new fountain in France can cause an old one in China to run dry.' Or again, should a king of Ethiopia decide to build a dam upstream of the Nile, 'perhaps you will see the whole of Africa's climate

19 Grove, *Green Imperialism*, p. 193.

20 Abbé Prévost, *Les Voyages du capitaine Robert Lade*, Paris: Didot, 1744, vol. 2, p. 93. Berta Pico, 'Les récits des voyageurs français aux Canaries: entre le mythe et la réalité. L'arbre saint de l'île de Fer', *Seuils et traverses: enjeux de l'écriture du voyage*, 1, 2002, pp. 79–88.

21 Lyon Municipal Library, MS Coste 1094, 'Mémoires d'un voyageur touchant les îles du détroit de la Sonde, Siam, la côte Coromandel, les Isles de France, quelques endroits de la côte d'Afrique'.

22 Louis-Bertrand Castel, *Esprit, saillies et singularités du P. Castel*, Amsterdam: Vincent, 1763, pp. 183–222. On Castel, see Jean-Olivier Richard, 'The Art of Making Rain and Fair Weather: The Life and World System of Louis-Bertrand Castel (1688–1757)', PhD dissertation, Johns Hopkins University, 2015.

change.' The impact of man's action on 'the organization and circulation of the whole globe' disrupted the water balance and hence the climate: 'What fogs, winds, rains, snows . . . floods, droughts, all of them accidental and only introduced into nature by the free action of human beings.' And Castel concluded that 'there may be no such thing as natural events'.[23]

It is therefore likely that Poivre drew inspiration from one or other of these sources. It is pointless to hypothesize mysterious Eastern influences – not least because he had nothing but contempt for animistic Chinese cosmology. One episode is particularly revealing. In 1749, while staying in Cochinchina (South Vietnam), he observed that gold, abundant in the mountains of the north, was not exploited in that country. He mocked the 'Cochinchinese idiots' who refused to extract it from the earth: 'Based on a mad belief [worship of the spirits of forests and rivers], the kings have forbidden that these invisible inhabitants be disturbed by felling trees, for fear of breaking the mysterious silence prevalent there, which their superstition adores.'[24] Like many intellectuals in the mid-eighteenth century, Poivre admired China. But he did so for the meticulous exploitation of its territory, certainly not for its forestry policy or philosophical systems. In short, as we can see, Grove's portrayal of a Poivre sensitized to environmental issues by his knowledge and love of China is unconvincing.

What, finally, of Poivre's posterity? What was the significance of his alert in the metropolitan climate debate? After his return to France, he took up residence in the château of la Fréta, not far from Lyon. He would never touch on climate matters again, except in his correspondence with his friend Jean-Nicolas Céré, to whom he entrusted the clove tree plantations on the Île de France. In Paris he frequented the salon of La Rochefoucauld, an aristocrat who was also a key figure in agronomic and philanthropic circles. We may assume that Poivre made certain French agronomists aware of the climate question. For example, in 1786 La Rochefoucauld read a dissertation by Céré on the climatic desiccation of the Île de France to the Société royale d'agriculture.[25] But it should be noted that Poivre's influence was slight: the naturalists, foresters and politicians who made climate change their warhorse from the 1800s onwards practically never mentioned him.

---

23 Castel, *Esprit, saillies et singularités*, p. 184.

24 A.N. Col C/1/2, f° 161–215, diary of a journey to Cochinchina.

25 Jean-Nicholas Céré, 'Mémoire sur la culture du riz à l'Île de France', *Mémoires d'agriculture, d'économie rurale et domestique publiés par la Société royale d'agriculture de Paris*, 1786, summer trimester, pp. 3–5.

Bernardin de Saint-Pierre, or An Unconditional Eulogy of Trees

Such influence as Poivre had was indirect. In fact, it was exercised especially, if not exclusively, through the writer and naturalist Bernardin de Saint-Pierre. Bernardin is infinitely more famous than Poivre, thanks to his exotic novel *Paul et Virginie*, one of the great bestsellers of the time.[26] This book, like the bulk of his work, was based on a two-year stay on the Île de France from 1768–70.[27] Bernardin was sent to the island in his capacity as an engineer to work on its fortification. At loggerheads with the military governor, he took refuge in the company of Pierre Poivre and his wife.[28] Poivre introduced him to botany: 'It is to him', wrote Bernardin, 'that I am indebted for the taste I have acquired for this field of study.'[29] There is no doubt that Bernardin made the intendant's climate theories his own. All his works would subsequently mention the influence of trees on rains and winds. This is true of *Paul et Virginie*, but also of his works of natural theology, which were almost as widely read and won him a certain recognition in the world of scholarship.[30]

It was in 1784, in his *Études de la nature*, that Bernardin de Saint-Pierre presented his theory of climate in depth. The book's aim was to refute atheism by unveiling the presence of a divine order identifiable in 'natural harmonies'. According to Bernardin, each element of nature must be interpreted in conjunction with the totality of Creation, of the Earth understood as one great organism. Hence his rejection of analytical thought that obliterated connections. Spinning a metaphor of nature as text, he mocked 'those who collect plants without registering their relations with one another and with the elements; they retain the letter, but delete its meaning.'[31] The clergy welcomed Bernardin as an antidote to

26 Martyn Lyons, 'Les best-sellers', in Roger Chartier and Henri-Jean Martin, eds, *Histoire de l'édition française*, Paris, 1990, pp. 409–88.

27 See *Voyage à l'Isle de France* (1773), *Études de la nature* (1784), and *Harmonies de la nature* (1815; posthumous). Jean-Michel Racault, Chantal Meure and Angélique Gigan, eds, *Bernardin de Saint-Pierre et l'océan Indien*, Paris: Classiques Garnier, 2011.

28 Letter to Hénin, 18 April 1770, *Correspondance de J.H. Bernardin de Saint-Pierre*, Paris: Ladvocat, libraire, 1821, vol. 1, p. 155.

29 Maurice Souriau, *Bernardin de Saint-Pierre d'après ses manuscrits*, Société française d'imprimerie et de librairie, Paris: 1905, p. 116.

30 After the Revolution, he was appointed intendant of the Natural History Museum.

31 Colas Duflo, 'Le finalisme esthétisant des *Études de la nature* de Bernardin de Saint-Pierre', in Catriona Seth and Eric Wauters, eds, *Autour de Bernardin de Saint-Pierre:*

Buffon and his materialism. His simple, natural religion inspired many sermons of the time.[32]

Chapter 11 of *Études de la nature* concerns the water cycle and the role played in it by mountains – a classic theme since Halley and Ray's writings on natural theology in the previous century. Bernardin inflected it with his engineering culture: mountains were like 'the hydraulic architecture of nature.'[33] But his true innovation lay elsewhere: his botanical observations convinced him that what trees bore, the form and texture of their leaves, was specific to each type of precipitation. For example, 'the pines of the sandy mountains of Norway collect the vapours that float in the air with their needles arranged like brushes.'[34] And he went much further: plants were not mere passive receptacles of celestial phenomena. Some plants, especially on mountains, had the capacity 'to *attract* the water that imbues the air in an imperceptible vapour' (our emphasis). To prove this hypothesis, he enlisted two clues. The first was none other than the sacred tree of El Hierro. Bernardin conceded that the phenomenon had 'possibly been somewhat exaggerated, but at bottom I believe it to be true.'[35] The second was drawn from his journey to the Île de France and to Bourbon, where he was able to observe how the rocky peaks attracted clouds before the trees caused them to empty themselves in rain.

Bernardin's theory of climate change was far more developed than Poivre's few and laconic allusions. It introduced three new features, decisive for the remainder of our history. Firstly, it stressed the role of *mountain forests*. It was the combination of mountains *and* trees that mattered: the first 'electrically' attracted the clouds which the second then condensed into rain.

Secondly, it saw the desiccation of the Île de France as revealing a much more serious phenomenon, affecting Europe and Asia in their entirety: 'I attribute to the same imprudence [the deforestation of mountains] the

les écrits et les hommes des Lumières à l'Empire, Rouen: Publication des universités de Rouen et du Havre, 2010, pp. 157–64.

32 Malcom Cook, *Bernardin de Saint-Pierre: A Life of Culture*, Abingdon: LEGENDA, 2006, p. 90; Maurice Souriau, *Bernardin de Saint-Pierre d'après ses manuscrits* [1905], Geneva: Slatkine, 1970, p. 228; Kurt Wiedemeier, *La Religion de Bernardin de Saint-Pierre*, Fribourg: Éditions universitaires de Fribourg, 1986, pp. 185–211.

33 Bernardin de Saint-Pierre, *Études de la nature*, vol. 2, Paris: Imprimerie de Monsieur, 1787, p. 324.

34 Ibid., p. 318.

35 Ibid., pp. 323–4.

palpable reduction of rivers and streams in much of Europe, as we can see by their former bed, which is much wider and deeper than the volume of water they contain today.' For Bernardin the island was a Creation in miniature, a small model of the world, where slow processes, imperceptible on the global scale, are speeded up.

Finally, and especially, Bernardin broke with the strictly 'desiccationist' or 'drying-up' conception of deforestation prevalent hitherto. In Buffon, for example, since trees humidify and cool the atmosphere, deforestation can be positive in a cold climate. According to Bernardin, on the contrary, trees are *always* beneficial and deforestation *always* an evil: 'forests offer shelter from the cold in the North, but what is admirable is that they offer shelter from the heat in hot countries.' To explain these seemingly contradictory effects, he once again referred to plant anatomy. In Northern countries, dominated by conifers, pine needles are 'shiny and glazed . . . and reflect the surrounding heat in a thousand ways: they generate the same effects as the coats of animals.' In hot countries, 'on the contrary, palms, talipots and coconut trees bear large leaves that on the ground side are matt rather than shiny.'[36] Whether fur coat or umbrella, trees were always beneficial.

Such an unconditional eulogy of the tree was something new. It is also fundamental for understanding the subsequent success of climate alarmism. With the forest no longer merely a cooling factor, but a *moderating* influence on climate – temperatures, humidity and winds – climate alerts could now batten on any kind of abnormal meteorological event.

An Energy Crisis

The influence of Bernardin's ideas does not proceed only from his talents as a writer or the support of the clergy. The publication of *Études de la nature* coincided with an intense energy crisis. The price of wood, on an upward curve since the 1770s, quadrupled during the winter of 1783–84. This crisis appeared to confirm predictions of the exhaustion of French forests. A scientific, prospective discourse warned of a coming shortage of wood throughout the kingdom. The government launched major national inquiries on forests, mines, and the industrial consumption of wood. The idea of planning a 'transition' to coal emerged.[37]

36 Ibid., vol. 1, 1784, p. 354.

37 Jérôme Buridant, 'Le premier choc énergétique: la crise forestière dans le nord

It was at this precise moment of anxiety about France's energy future that Bernardin's climate alarmism was diffused in academic circles and learned societies. In January 1786, the Société littéraire de Grenoble decided to devote its very first competition to the 'decline of woods'. Several essays already employed Bernardin's theses: deforestation aggravated storms – because the wind was no longer 'deadened' – as well as droughts.[38] The climatic argument also served to caution against the illusions, the dangers even, of a 'transition' to coal. In fact, according to some – Condorcet, for example – the benefit of ground coal was not that it preserved forests, but on the contrary that it could reduce their surface areas, as firewood became less useful.[39] This was an argument André Thouin inveighed against. A naturalist in the king's garden, he knew both Poivre and Bernardin. Coal, he reasoned, could certainly remedy the scarcity of wood, but certainly not the environmental disasters caused by deforestation: soil erosion, the loss of springs, and the disappearance of rain. He proposed a vast project of reforestation of *mountains* (the influence of Bernardin remains visible) aiming to restore the climates and hydrology of valleys.[40]

The Insular Origins of Collapse

On one side, greenness: the Dominican Republic, covered in dense forests. On the other, ochres, yellows and reds: Haiti deforested, desiccated, ravaged. The boundary seen from the sky as a sinuous line delimiting two ranges of colour. This photograph published in 1987 by *National Geographic* occasioned the spilling of much ink. It provided material for countless sombre commentaries on demography, poverty and deforestation, with collapse in the offing. In his book of that title, Jared Diamond devoted an entire chapter to this photo. In *Collapse* we find the amount

du bassin parisien début XVIIIe–début XIXe siècle', HDR thesis, Paris-IV, 2008; Reynald Abad, 'L'Ancien Régime à la recherche d'une transition énergétique. La France du XVIIIe siècle face au bois', in Yves Bouvier and Léonard Labories, eds, *L'Europe en transitions. Énergie, mobilité, communications, XVIIIe–XX1^{e} siècles*, Paris: Nouveau Monde Éditions, 2017.

38 *Mémoires de la société littéraire de Grenoble*, Grenoble: Allier, 1787, pp. 100–02.

39 Nicolas de Condorcet, 'Éloge de M. Morand', *Histoire de l'Académie royale des sciences, année 1784*, Paris: Imprimerie royale, 1787, pp. 48–53.

40 André Thouin, 'Mémoire sur les avantages de la culture des arbres étrangers pour l'emploi de plusieurs terrains de différentes natures abandonnés comme stériles', *Mémoires de la Société royale d'agriculture*, Winter 1786, p. 44.

Bibliothèque municipale de Lyon, Ms 5568 (collection of the Société royale d'agriculture de Lyon)

Prize competition of the Société royale d'agriculture de Lyon on the construction of effective chimneys, 1784. This frontispiece seems to link the competition's question to the forestry question and the equilibrium of the globe.

of Haitian forest cover cited as 1 per cent, even though some geographers have recently estimated it at 32 per cent, nearly as much as in France, and twice as much as in the United Kingdom.[41] That Haiti, the first Black colony to break its chains, should also be, in American neo-Malthusian discourse, an environmental catastrophe, is hardly surprising. But this cliché is actually older than Malthusianism. It originated in the history recounted in this chapter. In 1784, an agronomist applied the doctrines of Poivre and Bernardin to Saint-Domingue. In sum, he explained that the plantations were deteriorating in the part of the island under French domination (the future Haiti) because they were causing and experiencing a disastrous climate change which the Spanish part, less deforested, was spared.[42]

41 Jared Diamond, *Collapse: How Societies Choose to Fail or Survive*, London: Penguin, 2005, chapter 11; Christopher Churches, Peter Wampler, Wanxiao Sun and Andrew Smith, 'Evaluation of Forest Cover Estimates for Haiti Using Supervised Classification of Landsat Data', *International Journal of Applied Earth Observation and Geoinformation*, vol. 30, 2014, pp. 203–16.

42 Denis-Bernard Quatremère-Disjonval, 'Essai sur les caractères qui distinguent les cotons des diverses parties du monde', *Collection de mémoires chimiques et physiques*, Paris: Didot, 1784, vol. 1, pp. 296–300.

We may also take the case of Easter Island and its environmental collapse. Once again, it involves an old cliché that has been around in neo-Malthusian literature for a long time. Archaeologists have demonstrated its falsity: the great palm trees that used to cover the island did not disappear under the axe of the Rapa Nui as they erected their enigmatic *moai* but, gradually, under the teeth of the rats they had introduced. As for the abrupt demographic decline, this involves a long history, as banal as it is tragic, of European colonization, bacterial shock, and enslavement.[43] The cliché of an environmental collapse caused by the island's inhabitants once again has its origin in Poivre and Bernardin. The first European visitors to the island – Jacob Roggeveen in 1722, Don Felipe González in 1770, and James Cook in 1774 – all remarked on the absence of large trees, but none of them regarded it as a calamity. Roggeveen described an island that was certainly deforested, but nevertheless very fertile, with potatoes, bananas, sugar, and many fruits. 'The soil is rich and the climate is good, so much so that it could be transformed into a terrestrial paradise', he wrote.[44] By contrast, when in April 1786 the French explorer Jean-François Lapérouse landed on Easter Island, he perceived a nature fallen into ruin: an island 'without salutary shade', a soil 'calcinated by the strength of the sun', 'a horrible desiccation that makes it well-nigh uninhabitable'. The cause? The inhabitants, who 'have had the imprudence to cut down the trees'.[45] Here we find an echo of the theories of Poivre and Bernardin, which Lapérouse knew very well as a result of having spent five years in Mauritius.

Two and a half centuries later, islands still occupy a very special place in environmental discourse. Whether through the tragic fate of Easter Island in Diamond, or the coming fate of the Maldives, islands have become a laboratory of reports, the environmental closed room serving as a prelude to the future of the planet. This environmental and literary

43 Paul Ehrlich, who visited as a tourist in the 1990s, even makes it 'the monument to overpopulation'. Cf. Paul R. Ehrlich, *Human Natures: Genes, Cultures, and the Human Prospect*, Washington: Island Press, Shearwater Books, 2000, p. 244. Cf. Terry Hunt, 'Rethinking Easter Island's Ecological Catastrophe', *Journal of Archeological Science*, vol. 34, no. 3, 2007, pp. 485–502. Cf. Patricia A. McAnany, *Questioning Collapse: Human Resilience, Ecological Vulnerability, and the Aftermath of Empire*, Cambridge: Cambridge University Press, 2010.

44 Bolton Glanvill Corney, ed., *The Voyage of Captain Don Felipe González to Easter Island 1770–1, preceded by an extract from the Official Log of Mynheer Jacob Roggeveen in 1722*, Cambridge: The Hakluyt Society, 1903, p. 21.

45 *Relation abrégée du voyage de La Pérouse, pendant les années 1785, 1786, 1787, et 1788*, Leipzig, 1799, pp. 59–62.

motif was invented in the late eighteenth century under the sway of Poivre and Bernardin. In a sense, the thesis of Green Environmentalism, the idea of the insular, colonial birth of ecology, is still a reflection – more academic and more subtle – of this *topos*. For, as this chapter has shown, the Île de France under Poivre's administration was less the nursery of the environmentalism described by Grove than the construction of an arsenal and the birthplace of a trope from the pen of Bernardin de Saint-Pierre: the island as the sentinel of collapse.

1789: revolution breaks out in Paris. From Saint-Domingue the colonists send a denunciation to the National Assembly demanding the dismissal of La Luzerne, colonies minister and former governor of the island. What is he accused of? Among other things, of having, by granting concessions in the mountains, 'completely changed the climate by the sudden destruction of forests'.[46] The rest of the indictment reads like a précis of the *Études de la nature*. Islands unquestionably played a part in the formation of the climate alert, but it was only during the Revolution, and as a result of it, that that alert transcended the bounds of scholarly circles as, in the white heat of revolutionary assemblies, climate became a battlefield, a way of blaming political opponents for the breakdown of nature.

46 Louis-Marthe de Gouy d'Arsy, *Première Dénonciation solennelle d'un ministre, faite à l'Assemblée nationale, en la personne du Comte de La Luzerne*, Paris: Demonville, 1790, pp. 121–31.

6

The Climate of Revolution

The French Revolution was also a climate revolution. During the revolutionary events, and as a result of them, climate warnings were issued by naturalist circles and became a political issue, dealt with in arenas of power, the National Assembly and ministerial offices. Scientists who occupied the summits of power, deputies, landowners, prestigious agronomists, and ministers were convinced of the reality of the threat.

How do we explain the fact that climate change was at the heart of political concerns? Circumstances certainly played a role: the terrible weather of 1788 and 1794 caused grain shortages and food riots. The increase in the price of wood and the needs of a war economy reinforced fears of a depletion of French forests. But this experience of natural limits, typical for a natural economy, was shared by many other regions in Western Europe. It is not enough to justify the particularities of the French climate debate.

Three major revolutionary challenges account for the precocious, radical politicization of the climate in France. First: how was a soil and a nation degraded by feudalism to be regenerated? And – a correlated question – what would the Revolution's consequence be for the nature of France? Second: what was to be done with the immense national forestry domain inherited from Church property? How should this gigantic capital be used to strengthen the state's credit, stabilize the *assignat* and finance wars? Finally, and especially: how were the revolutionary rural masses to be governed? How to inculcate respect for forests, but also for property and order, in a population of peasants who had become citizens?

'Repairing the Climate'

'Is there a way of repairing a country, a climate?'[1] This question raised in 1790 perfectly illustrates the way that revolutionary discourse seized on the climate: at once as a means of regeneration and, reparation following damage, as an indictment of the past. The originality of the French climate debate resides in its link with political confrontation. From 1789 to the 1820s, every regime claimed to clear up the climatic mess created by the negligence of its predecessors. At the start of the Revolution, climatic discourse served to denounce the environmental effects of despotism and feudalism. According to the Jacobin deputy Jacques-Michel Coupé, climatic imbalance was attributable to 'the prolonged negligence [that] reigned over French territory. . . . There is a great deal to *repair* on its surface.'[2]

If feudalism had debased even the French climate, it was (according to the revolutionaries) because it was not only unjust, but unnatural. It not only fettered a potentially generous nature, it destroyed it. Two massive catastrophes supplied proof of this: the proliferation of swamps and pools on the one hand, deforestation on the other.

The revolutionary offensive against stagnant water – with the drainage decrees of 26 December 1790 for marshes and of 4 December 1793 for ponds – drew on older medical theories and analyses.[3] Post-1789, however, the charge assumed extraordinary gravity: by allowing water to stagnate, the monarchy had degraded the French climate and people alike. The French, 'issued from the same Gallic mould',[4] now exhibited various degrees of physical degeneration. Marshes and ponds not only generated bad air, they also altered the climate in the meteorological sense. According to a Jacobin deputy, they produced 'torrential downpours' and

1 Pierre-François Boncerf, *Mémoire sur les moyens de mettre en culture les terres incultes, arides et stériles de la Champagne*, Paris: Gorsas, 1790.

2 Archives Nationales, AN AD/IV/19 Jacques-Michel Coupé, *De l'amélioration générale du sol français, dans ses parties négligées ou dégradées*, [publisher and date unknown], p. 1.

3 Reynald Abad, *La Conjuration contre les carpes*, Paris: Fayard, 2006 and Jean-Michel Derex, 'Le dessèchement des étangs et des marais dans le débat politique et social français du milieu du XVIIIe siècle à la Révolution', in Salvatore Ciriacono, ed., *Eau et développement dans l'Europe moderne*, Paris: MSH, 2004, pp. 231–47.

4 Jacques-Antoine Boudin, *Mémoire sur le dessèchement et la mise en culture des étangs, de la Sologne, de la Bresse, de la Brenne*, Paris: Convention nationale, 22 Brumaire Year II, 1793, p. 14.

thus became 'sources of hail and late frosts'.[5] The seigneurial mills with their reservoirs and fish ponds – a privilege of monastic orders and feudal lords – fell within the scope of the same climatic indictment.[6] In 1790, the Royal Agricultural Society even proposed replacing hydraulic mills by hand mills or carousels with horses to preserve local climates.[7] Against a feudal regime that blocked waterways, multiplied swamps, left plants to rot and the population to degenerate, revolutionary discourse pitted a republic that made things flow, drained and cultivated, warmed climates and rendered them healthy.

Deforestation was the second major harm done by despotism to the French climate. This charge, too, was extremely serious. The people were cold; the navy, forges and workshops lacked the wood required to defend the nation; humus was sliding off deforested slopes causing erosion, torrents, and floods; depleted mountains were not stopping storms; clouds passed without turning into rain. Hence the recurrence of droughts, hurricanes, hailstorms and frosts. The charge was all the weightier in that it followed a series of meteorological disasters. The drought of summer 1788 was followed by torrential rains. There ensued the heavy winter of 1788–89 (eighty-six days of ice in Paris). The price of bread doubled in the capital and the popular classes spent three-quarters of their income on it.[8] Lists of grievances resounded with complaints against the erratic climate, often accompanying demands for tax rebates.[9] The food issue that lay in the background of the Revolution was construed as a climatic event with political causes.[10]

Once again, the charge facilitated a critique of the very foundations of

5 Ibid. and Marie-Émilie-Guillaume Duchosal, *Discours sur la nécessité de dessécher les marais, de supprimer les étangs et de replanter les forêts, prononcé le 12 mai 1791 dans la Société patriotique de la section de la Bibliothèque*, 1791, p. 9.

6 Claude-Louis Berthollet et al., *Rapport général sur les étangs*, Paris: Imprimerie de la Feuille du Cultivateur, 5 Nivôse Year III, p. 121.

7 Valmont de Bomare and Pierre-François Boncerf, *Moulins à bras et à manège*, Paris: Valade, 24 August 1790.

8 Le Roy Ladurie, *Histoire humaine et comparée du climat*, vol. 2, pp. 143–80.

9 In Provence, in particular, people bemoaned frosts and hailstorms that had become more frequent since the 1770s. Cf. *Archives parlementaires de 1787 à 1860*, vol. 3, p. 446, Sénéchaussée d'Aix.

10 The storm of 13 July 1788 was thus meticulously studied by the Académie des sciences, which estimated the loss to agriculture at 25 million livres (5 per cent of the state budget). Cf. Leroi, Buache and Tessier, 'Second mémoire sur l'orage de grêle du 13 juillet 1788', *Histoire et Mémoires de l'Académie des sciences*, pp. 263–85.

the ancien régime. Firstly, the water and forestry authorities were said to be riddled with venality around the sale of offices, a source of incompetence and corruption.[11] Secondly, the logic of a personal reign and quest for military glory were incompatible with long-term management of forests: Louis XIV had 'eaten up the resources of other generations for the greater glory of the time of his reign'.[12] Thirdly, by fostering the taste for luxury, the court had increased the national consumption of wood. Lists of grievances demanded a return to sumptuary laws or taxes in order to put an end to 'the superfluous consumption of wood' by nobles and bourgeois.[13] Added to this was the obsession of Physiocratic economists, counsellors of Louis XV and Louis XVI, with grain. 'Their *economistic* frenzy for land clearance' had razed forests, 'those physical instruments for maintaining the country's temperature'.[14] It had even set off a vicious circle, at once Malthusian and climatic: with each grain crisis, the government encouraged deforestation; thanks to the expansion of the cultivated area, the population increased, but so did climatic risk and hence the threat of shortages, which inevitably followed, prompting further deforestation, and so on.[15]

'Compelling Time to Release Its Prey'

The idea of French nature being on the brink of the abyss made it possible both to incriminate the past and to glorify the Revolution's restorative work. The key theme of 'regeneration'[16] thus had plenty of environmental

11 The lists of grievances were bursting with recriminations against supervisors of waters and forests – and demands for the communal management of woods. See Peter McPhee, '"The Misguided Greed of Peasants"? Popular Attitudes to the Environment in the Revolution of 1789', *French Historical Studies*, vol. 24, no. 2, 2001, pp. 247–69; Kieko Matteson, *Forests in Revolutionary France: Conservation, Community and Conflict*, Cambridge: Cambridge University Press, 2015.

12 Rougier de La Bergerie, *Mémoire et Observations sur l'abus des défrichements avec un projet d'organisation forestière*, Auxerre: Fournier, Year IX, 1800, p. 7.

13 *Archives parlementaires de 1787 à 1860*, Paris: Librairie administrative, vol. 2, p. 243 (bailiwick of Bar-le-Duc).

14 AN AD/IV/19, Coupé, *De l'amélioration générale du sol français*.

15 La Bergerie, *Mémoire et observations sur l'abus des défrichements*. A decree of 1766, exempting newly cultivated land from taxation, was in the sights of revolutionary foresters. No less than 359,000 acres of woods were said to have disappeared between 1766 and 1774, during the great wave of land clearance launched by the Marquis de Turbilly.

16 Mona Ozouf, 'Régeneration', in François Furet and Mona Ozouf, eds, *Dictionnaire critique de la Révolution française*, Paris: Flammarion, 1988, pp. 821–31 and Emma Spary,

applications. A forester explained that 'it is not enough to have regenerated the Frenchman, it is necessary to regenerate his land'.[17] One of the most striking aspects of revolutionary climate discourse is its radical historicism. Steeped in the vision of a human history of nature (Chapter 3), the revolutionaries regarded the environment of France as the product of contingent processes, under the influence of a bad political regime. For example, according to the deputies Coupé and Boudin, promoters of the drainage policy, *all* stagnant waters were the work of man: swamps were simply older ponds whose origin was lost in the mists of time.[18] Similarly, an absence of forests had to be the sign of previous deforestation. This process was more or less advanced according to the antiquity of the civilization. Its endpoint was the Middle East, where civilization was born and where, on account of growing aridity, it was being extinguished. According to Coupé, the Sahara itself was the result of oriental despotism. Greece and Spain, like the Cévennes in France, were already far advanced on the path of decline. According to the chemist Berthollet, what awaited France was the 'burning deserts of Asia, of this Greece . . . which would still enjoy a favourable temperature if the most barbaric tyranny and ignorance had not resulted in the disappearance of the woods'.[19]

Happily, the Revolution betokened a fork in the road. A total event, it embraced nature in its entirety. Becoming conscious of itself, popular sovereignty would be capable of realizing Buffon's dream of climate control; it would 'show a state giving its power over nature a first try'.[20] According to François-Antoine Rauch, an engineer with Ponts et chaussées to whom we shall return (Chapter 10), 'the Frenchman is going to feel his power'.[21] 'France', he went on, could become 'the earthly paradise of Europe', because it 'alone has as many scientists as the rest of this part of the globe'. 'Such a lofty mission [regenerating, acclimatizing, etc.] would

Le Jardin de l'Utopie: l'histoire naturelle en France entre Ancien Régime et Révolution, Paris: Muséum national d'histoire naturelle, 2005.

17 AN AD/IV/20 Bourlet, *Précis sur la nécessité d'organiser promptement l'administration forestière*, [n.p.:n.d.], p. 3.

18 Jacques-Antoine Boudin, *Du dessèchement des marais et terreins submergés*, Paris: Imprimerie nationale, Comité de salut public, [n.p:n.d.], p. 27.

19 Berthollet et al., *Rapport général sur les étangs*, p. 120.

20 Pierre Laureau de Saint-André, 'Destruction générale de la mendicité dans toute la France', *Archives parlementaires*, 21 January 1792, vol. 37, pp. 566–70.

21 François-Antoine Rauch, *Plan nourricier, ou Recherches sur les moyens à mettre en usage pour assurer à jamais le pain au peuple français*, Paris: Didot, 1792, p. 20.

transform our scientists into demi-gods and render science tangible and precious to humanity.'[22]

If climatic regeneration was guided by 'semi-divine' scientists, in practice it had to be effected with shovels and pickaxes by poor people, whom it would usefully occupy. Pierre-François Boncerf, whose idea of 'climate repair' opened this chapter, was a renowned agronomist, a supervisor of public works in Paris.[23] It was in the latter capacity that he proposed to the National Assembly that it hire the workers of Paris, made unemployed by the Revolution, to drain and reforest France. The advantage was twofold: on the one hand, 'repairing the climate'; on the other, removing from Paris idle workers who posed a threat to public order. Boncerf thought big, very big: the national territory must be cut up into squares of 10,000 acres which it would be necessary to irrigate, drain, or reforest. The process would be iterative and progressive, with each forest planted improving the climate of the adjacent parcel. The goal of these gigantic works was to 'change the climate, restore the commerce between Earth and heavens, set meteors into motion again and multiply them'.[24]

The plan was taken up by the deputy Pierre Laureau de Saint André. Now it was brigades of foundlings and beggars that were to be sent into the mountains to reforest and terrace them in order to prevent erosion, restore the rains and the climate.[25] The Revolution was set to repair the damage of feudalism and, what was more, arrest even the effect of time that eroded nature: 'It is a great truth that man wears down the soil . . . in his wake nature is like sand trickling away in the hand of time. I propose to pit against the enterprises of time on our soil the arms of beggars.' The Revolution will save France from the fate of Egypt and its deserts. It will compel 'time to release its prey'.[26]

---

22 Ibid., p. 91.

23 Pierre-François Boncerf, *Précis de la défense du citoyen Boncerf au tribunal révolutionnaire*, 18 Ventôse Year II, p. 2. His pamphlet of 1776 on *Inconvénients des droits féodaux* is said to have inspired the decrees of the night of 4 August.

24 Boncerf, *Mémoire sur les moyens*, 1790, p. 8.

25 Laureau de Saint André was also a historian. His *Histoire de la France avant Clovis* adopted the historico-climatic circle described in Chapter 3. According to him, the word 'Celt' signified 'cold climate' in that language. Cf. *Histoire de France avant Clovis*, Paris: Nyon, vol. 1, 1789, pp. 7, 30–7.

26 Saint André, 'Destruction générale de la mendicité'.

'The Forestry Security'

On 2 November 1789, the Constituent Assembly voted the nationalization of Church property, which included immense wooded tracts estimated to cover more than one million hectares.[27] In the spring of 1792, debates in the National Assembly were heated: should these woods be sold, to rid the state of its debts and arm France against its enemies? The deputies' response was negative: they accepted the privatization of 'copses' of less than 100 acres (around 40 hectares),[28] monitoring of which was hardly profitable, but large expanses of forest were declared inalienable.

However, in strictly accounting terms, the benefits of privatization seemed unassailable. According to the deputy from the Jura, Simon Vuillier, the national forests earned the state seven times less than their rental value. It was also an easy asset to sell: the rising price of wood guaranteed its value. To these arguments, Vuillier added another: the authorities were poor managers of forests, which required the particular care only an owner could provide. He even stated that he would prefer the free distribution of national forests among the citizenry to leaving them under the control of incompetent civil servants.[29]

Vuillier's speech to deputies (on 2 March 1792) caused uproar. He was accused of being the accomplice of 'black companies' of financiers eager to get their hands on French forests, even of being a foreign agent. The coming war (it was declared on 20 April) conferred a strategic value on forests: they would supply the needs of the fleet and serve as 'security', enabling debts to be incurred to meet military expenditure. One question above all troubled the lawmakers: what would become of the *assignats* if the national forests were sold? They were already afraid that, with the fall in their value, the European powers could buy the wood required to fight

27 260,000 buildings, 4.7 million hectares – at a conservative estimate, 10 per cent of the national heritage. Cf. Bernard Bodinier and Éric Teyssier, *L'Événement le plus important de la Révolution: la vente des biens nationaux en France et dans les territoires annexes: 1789–1867*, Paris: CTHS, 2000. With the royal forests, the national forestry estate amounted to two million hectares. Cf. *Archives parlementaires*, vol. 40, p. 721 and *L'Administration des eaux et forêts du XIIe au XXe siècle*, Paris: CNRS, 1987.

28 AN AD IV 19 'Décret de l'Assemblée nationale du 6 août 1790 qui excepte les grandes masses de bois et forêts de l'aliénation des biens nationaux'. We might note that the Directory increased the limit to 300 acres (law of 2 Nivôse Year IV).

29 *Archives parlementaires*, vol. 39, p. 303. Vuillier would profit from the sale of national assets to carve out an estate for himself near Dôle. Cf. Claude-Isabelle Brelot, *Grands notables du Premier Empire. Jura, Haute-Saône, Doubs*, Paris: CNRS, 1979, p. 57.

France for a song. In addition, the *assignat* of 1789 was peculiar in being based not on stocks of precious metals, but on the *territorial wealth* of the nation. By making forests a special, inalienable national asset, deputies sought to ensure it an impregnable underpinning.[30]

Along with the financial issues decisive for the refusal to sell, the fate of the national woods also elicited questions about the future of France, its economy and climate. For example, should one trust the promises of coal? According to the advocates of selling, the fears of a wood shortage expressed in the Assembly were irrelevant, because coal was about to replace it. To which opponents retorted: 'coal and peat are neither renewable nor abundant, they are talismans that fascinate minds.'[31] The Assembly's Agriculture Committee confirmed that in the long run, wood would remain the principal fuel, because coal mines 'are not as common as is thought. We see that those in the Auvergne are becoming exhausted.'[32] In 1792 the exploitation of coal seemed to many a merely temporary solution, a stopgap while awaiting the reconstruction of France's forests.

It was also during these debates that the climate change argument made its appearance in the National Assembly. The Royal Agricultural Society warned deputies against the climatic effects of deforestation, which were already making themselves felt in Provence.[33] A deputy opined that the sale of the national forests was a first step in the complete privatization of nature: 'It will soon be proposed to you that we sell our streams, our rivers,

30 The *assignat* was supposed to make it possible to resolve the paradox that the French state was unable to pay its creditors even though it was sitting on an enormous mass of real estate. Alain Alcouffe, 'Vandermonde, la monnaie et la politique monétaire de la Révolution', *Annales historiques de la Révolution française*, no. 273, 1988, pp. 254–64; Michel Bruguière, 'Assignat', in François Furet and Mona Ozouf, eds, *Dictionnaire critique de la Révolution française. Institutions et creations*, Paris: Flammarion, 1992; Rebecca Spang, *Stuff and Money in the Time of the French Revolution*, Cambridge, MA: Harvard University Press, 2015; Jean-Luc Chappey and Julien Vincent, 'A Republican Ecology? Citizenship, Nature and the French Revolution (1795–1799)', *Past and Present*, vol. 243, no. 1, 2019, pp. 109–40.

31 *Archives parlementaires*, vol. 40, p. 748.

32 Ibid., vol. 39, p. 292, 'Rapport des comités réunis sur la question de la vente des forêts nationales'.

33 *Observations sur la question de l'aliénation des forêts nationales présentées à l'assemblée nationale par la Société royale d'agriculture le 3 février 1792*, Paris: La Feuille du cultivateur, 1792. The Agricultural Society initially inclined to privatization. Cf. *Observations sur l'aménagement des forêts et particulièrement des forêts nationales, présentées à l'assemblée nationale par la société royale d'agriculture le 9 juin 1791*, Paris: La Feuille du cultivateur, 1791.

our brooks, even the air we breathe. Not enough thought has been given to the fact that the purity of the air, the very existence of water, is due to our forests.'[34] Above all, it was during this debate that two key figures in the climatic and forestry struggles of the early nineteenth century spoke up on these issues for the first time: the engineer François-Antoine Rauch and Jean-Baptiste Rougier. The latter was a prominent figure, a wealthy property-owner in the Yonne, a deputy and member of the Royal Agricultural Society. In March 1792 he issued a warning that went so far as to present the fate of the national forests as the tipping point in the regeneration or decline of France. The history of Egypt, Persia and Arabia must put the nation's representatives on their guard: 'millions of men have died with the exhaustion and desiccation of the land'.[35]

The nationalization of Church forests was a turning-point in our history, for it yoked the issue of climate change to that of state credit. The National Assembly long resounded with the echoes of the debate of March 1792, when deputies rejected privatization for reasons that were in part climatic.

'Stop, Stop That Fatal Axe!'

When, in 1789, the revolutionary deputies accused the monarchy of having degraded the French climate, perhaps they could imagine how this accusation would rebound against them. One month after the night of 4 August, the environmental consequences of the abolition of privileges were already being discussed in the Assembly. Deputies were distraught at the liberties taken by peasants, poaching, fishing and, above all, committing 'misappropriation of the woods'. The discourse of collapse switched targets: the point was no longer to accuse the ancien régime, but to seek to control the rural masses in their relationship with nature.

From summer 1789, a flood of police reports, statements and petitions submerged the Assembly, particularly its Agriculture Committee, warning of a disaster unfolding in the forests. These reports and complaints

34 *Opinion de L.C. Chéron sur les dangers de l'aliénation des forêts nationales*, Paris: Imprimerie nationale, 1792.

35 Rougier nevertheless declared himself *in favour* of a sale, because private parties 'would employ thousands of hands' to maintain the forests. The forests must not be sold to big companies, but divided up into small parcels so that they could be acquired by citizens with little wealth who, thanks to the woods, would be connected with the nature of France. Cf. *Archives parlementaires*, vol. 39, p. 310.

described generalized rapine, the theft of wood by gangs, wild clearances, including on the slopes of mountains, or mass grazing in national woods. The environmental argument made it possible to raise the stakes: it was a question not merely of theft and loss of income to an owner, but of erosion and climate change. In the Gard, the vineyards seemed doomed to sterility because of the cold and wind caused by the deforestation of hillsides. A petition signed by 300 citizens of Béziers blamed the freezing of olive trees on deforestation, which allowed the north winds to batter the Mediterranean. In Lozère, it was the clearing of the mountains of Auvergne that had supposedly caused the sweet chestnut trees of the Cévennes to wither, and so on.[36] These alarmist discourses compiled by Rougier de La Bergerie would give rise to the black legend surrounding the Revolution: the 'devastation of woods'. Michelet delineated it in some famous pages: 'with the Revolution, every barrier fell: the poor began the work of destruction. . . . Trees were sacrificed for the smallest purposes; two pines might be felled to make a pair of clogs.'[37]

Why did peasants target trees? Or, rather, why had certain trees become emblematic of seigneurial power, as had large ponds?[38] To understand this, we need to go back to the 1760s when, in a context of rising wood prices, lords and large property-owners constructed forestry estates of a new kind, given over to markets in softwood lumber or wood charcoal and exempt from user rights. The nature of forests changed: open spaces fashioned by silvopasture were replaced by forestry monocultures defended by guards

36 La Bergerie, *Mémoire et observations sur l'abus des défrichements*, p. 30.

37 Quoted and discussed in Denis Woronoff, 'La "dévastation révolutionnaire" des forêts', in Woronoff, ed., *Révolution et espaces forestiers*, Paris: L'Harmattan, 1988, pp. 44–52. Measuring the impact of the Revolution on forests is impossible: we would need to know about clearances in private woods that were no longer subject to control from 1791. All assessments of the wooded surface area before 1830 are far-fetched because they significantly underestimate private woods. Cf. B. Cinotti, 'Évolution des surfaces boisées en France: proposition de reconstitution depuis le début du XIXe siècle', *Revue forestière française*, vol. 48, no. 6, pp. 547–62. Instead, we need to locate the development of forest cover under the revolution within a long-term dynamic extending from the sixteenth to the mid-nineteen century. According to Buridant, the period 1780–1840 corresponds to the low point of forestry surfaces, but the total surface area did not decline. Cf. Buridant, 'Le premier choc énergétique'.

38 This whole section owes much to Florence Gauthier, *La Voie paysanne dans la Révolution française. L'exemple de la Picardie*, Paris: Maspero, 1977 and Anatoli Ado, *Paysans en révolution. Terre, pouvoir et jacquerie 1789–1794*, Paris: Société des études robespierristes, 1996.

and often ringed with walls. Various feudal rights allowed the lords to abolish communal user rights or to appropriate part of the common land on condition of planting it with trees.[39] The result was that, by the late eighteenth century, they had monopolized wood resources in numerous regions.[40] This privatization of forests was a formidable attack on peasant ways of life, given the vital importance of access to woods in order to feed animals, obtain firewood, fashion tools, gather manure, and so on.

From July 1789, peasants throughout France took cognizance of the Revolution by restoring user rights over forests. They expelled the forest guards, hunted game, gathered wood, and sent their livestock to graze, especially in the ecclesiastical and royal forests that had become 'national property'. A response to concrete needs, this also involved an active struggle against the usurpation of their prerogatives and common land. This reaction proceeded from political acts that were sometimes solemnly and collectively carried out, with the felling of trees in processions directed by the municipal council.[41]

The problem was that the deputies of 1789 were bitterly opposed to restoring user rights; on the contrary, they defended full, outright ownership of forest spaces. The people were said to have misunderstood the idea of 'nationalization'. In the language of parliamentarians, old user rights became forestry crimes, brigandage, pillage. Following the night of 4 August 1789 that represented its great victory, the peasant revolution clashed head-on with the new government. A decree of 11 December 1789 forbade communities to 'take possession of any woods, pasture, waste land of which they did not have possession on 4 August last'. In 1791 a 'rural criminal law' provided for severe penalties – up to one year in prison – for stealing wood. Grazing in national forests was punished by heavy fines. This legislation that criminalized usages rendered them furtive, unregulated, and so more harmful.[42]

39 Matteson, *Forests in Revolutionary France*; François Vion-Delphin, 'Forêts et cahiers de doléances: l'exemple de la Franche-Comté', in Woronoff, ed., *Révolution et espaces forestiers*, pp. 11–22; Marcel Dorigny, 'Usages forestiers communautaires et demandes protoindustrielles dans l'Autunois', in ibid., pp. 28–36.

40 Not less than 40 per cent in Picardie. In this province, communes that still had forests were rare. Cf. Gauthier, *La Voie paysanne dans la Révolution française*, pp. 78–81.

41 Cf. Ado, *Paysans en révolution*, p. 211ff.

42 AN AD IV 19 Laws of 4 November 1790, 19 December 1790, 27 December 1790, 22 April 1791. Cf. *La Feuille villageoise*, vol. 2, no. 46, 11 August 1791, p. 367.

Troops were sometimes sent to keep guard in forests. In the Aude, studied by Peter McPhee, there were violent clashes.[43] In Montereau, some peasants refused to disperse when confronted with national guards, 'asserting that they were on their own land, were cutting down trees that belonged to them, and had the right to do so'.[44] Even though the deputies relied on property rights for the purposes of conservation, it was precisely that full and outright ownership of forestry spaces that seemed intolerable to the peasants, as intolerable as the privileges abolished on the night of 4 August. When a property-owner in the Nièvre threatened to prosecute neighbours who were stocking up in his woods, they retorted: 'Do you think we made the revolution for nothing?'[45]

What is the link between the peasant revolution and climate change? Although repression was severe, with prison sentences often handed down against 'ringleaders', the widely shared assessment was that control and repression would not suffice to protect forests. Forestry guards could not prevent the collection of wood, which was as massive as it was common.[46]

In addition, after the very important forestry law of 28 September 1791, the 4 million hectares of private forests (at a conservative estimate) out of a national total of 7 million were no longer subject to *any regulation*.[47] Owners were henceforth free to manage their wood 'as they saw fit', to decide on tree felling and even, if they so desired, to clear the land.

So, there was no way out: regardless of who owned the forests, the whole population had to be persuaded to respect them. And it was therefore necessary to operate indirectly, through education, instilling in citizens the idea that trees played an essential role in the natural and providential order. Hence the emergence of a semi-scientific discourse, addressed to the peasant masses and landowners, on the crucial climatic importance of forests.

In the autumn of 1794, the harvests were set to be catastrophic.[48] During the summer, storms and hail destroyed crops. Compounding this

43 Peter McPhee, *Revolution and Environment in Southern France, 1780–1830: Peasants, Lords, and Murder in the Corbières*, Oxford: Oxford University Press, 1999.

44 Quoted in Ado, *Paysans en révolution*, p. 211.

45 AN F10 403.

46 We find many examples of police interventions in the forests in the police archives of the revolutionary period AN F7 3647 (Ain) to AN F7 3686 (Haut-Rhin).

47 Marie-Noëlle Grand-Mesnil, 'La loi du 29 septembre 1791', in Woronoff, ed., *Révolution et espaces forestiers*, pp. 200–05.

48 Le Roy Ladurie, *Histoire humaine et comparée du climat*, vol. 2, pp. 194–234.

were the naval blockade, military requisitions, and the refusal of large-scale farmers to sell their grain at the price fixed by the state. The new Thermidorian government confronted food riots. Such was the turbulent context in which Rougier de La Bergerie was sent into the Creuse. A member of the Agriculture Committee, he was tasked with assessing the scale of the damage caused by a hailstorm that had hammered the department. His tour also afforded an opportunity to give 'fraternal agricultural lectures', which in truth were distinctly unfraternal. Rougier explained to the doubtless astounded peasants that they were *responsible for their own climatic misfortunes*: 'You owe . . . the frequency of bad weather and storms to the devastation of your woods.'[49] The guilt-inducing role of the climate is clear: 'Let us pit all our efforts against these unfortunate *effects*, which are distressing, but which we *deserve* for our lack of foresight in clearing our mountain woods'.[50] Climatic disasters were no longer a cause, but an *effect* – an effect of peasant greed and ignorance. Misled by a false idea of liberty, country folk thought they could emancipate themselves from forestry rules that were also natural rules: 'an ever beneficent, ever wise nature had covered your highest mountains with trees and therefore these were necessary.' Rougier ended with an injunction: 'Stop, stop that fatal axe.'[51] In the midst of a food shortage, when the peasants were opposing military requisitions, climate change made it possible to shift responsibility for hunger: the victims were their own executioners.

The same year, the Abbé Grégoire – a major figure in the Revolution who was also an experienced agronomist – proposed to the Convention that it establish in each district 'a centre of rural economy' tasked with teaching peasants modern agricultural methods.[52] These agricultural schools were also to impart a respect for trees, for, he wrote, they 'stop the clouds and resolve them into fertilizing rains'.[53] Though his plan for agricultural schools did not come to fruition, Grégoire actively contributed to

49 Rougier de La Bergerie, *Traité d'agriculture pratique, ou Annuaire des cultivateurs du département de la Creuse et pays circonvoisins, avec des vues générales sur l'économie rurale*, Paris: 1795, p. 378.

50 Ibid., p. 97.

51 Ibid.

52 Jean Boulaine, 'La carrière agronomique de l'abbé Grégoire', *Comptes rendus hebdomadaires des séances de l'Académie d'agriculture de France*, 1990, vol. 76, no. 1, pp. 83–9.

53 *Rapport et Projet de décret, sur les moyens d'améliorer l'agriculture en France par l'établissement d'une maison d'économie rural dans chaque département*, Paris: Convention nationale, 1793, pp. 4–5.

the development of a republican culture of trees. Initially, the Revolution had erected maypoles, flagpoles or trees whose trunk was intact, simply stuck in the ground: reaching skywards, their silhouette recalled the scaffold. Symbols of seigneurial privileges were affixed to them and they represented the peasant revolution. But what triumphed with Grégoire or Rougier was different: the celebration of 'liberty trees', wisely planted, that fuelled an institutional republican imagery stressing the regime's longevity and fertility.[54]

In 1797, under the Directory, the tradition of liberty trees turned into a government programme for the creation of nurseries intended, among other things, to restore the climates of France. François de Neufchâteau, interior minister and agronomist, introduced bonuses to encourage planting. 'We have restricted ourselves', his circular explained, 'to planting one Liberty Tree in each commune. A single tree is sad. Let us sow the seeds of whole woods; let us plant vast forests; let us raise natural temples to Liberty.' Republican sylviculture must instil a veritable love of trees in citizens (especially children). And to that end, local authorities must wield the climate threat: 'show the influence of major plantations on variations in the atmosphere [and] how many communes whose soil [has been] sterilised by prolonged droughts'.[55]

The climate warning was also relayed by *La Feuille du cultivateur* and *La Feuille villageoise*, two papers that carried the government's words into the countryside. Their articles seem to touch on a wide range of subjects, from agronomy to new policies via the climate disorders created by deforestation.[56] But they generally had a precise goal: 'to enlighten the peasant emboldened by revolution'[57] (such was the slogan of *La Feuille villageoise*); to influence the behaviour of rural folk so as to render agrarian individualism compatible with national regeneration. At bottom,

54 Mona Ozouf, 'Du mai de liberté à l'arbre de la liberté: symbolisme révolutionnaire et tradition paysanne', *Ethnologie française*, vol. 5, 1975, pp. 9–32.

55 AN F10 276, Circular of 22 Fructidor Year V.

56 For example, Philippe Bertrand, 'Mémoire important sur l'économie politico-rurale des pays de montagnes', *Introduction à "La Feuille du cultivateur"*, Paris: *La Feuille du cultivateur*, 1793, pp. 3–6; Céré, 'Mémoire sur la culture du riz à l'Île de France par le C. Céré, communiqué par Larochefoucauld', in ibid., p. 34. Rougier de La Bergerie, 'De l'utilité de garner d'arbres les sommets des montagnes', *La Feuille du cultivateur*, vol. 5, pp. 105–10. 'Sur la nécessité de multiplier les arbres en France', *La Feuille villageoise*, vol. 5, no. 50, p. 210.

57 The prospectus invited 'patriotic priests' to assemble the peasants each week to read it to them. Cf. *La Feuille villageoise*, vol. 1, p. 8.

climatic discourse was one aspect of the more general project of restoring order to the countryside after the great peasant revolt of 1789. 'We have triumphed over all tyrannies, we must now banish all storms and bad weather', explained Rougier de La Bergerie.[58] Here, as elsewhere, what was at stake was bringing the Revolution to a close.

Napoleon and the Water Cycle

After 18 Brumaire, this programme assumed a special relevance: Bonaparte's arrival was supposed to end rural unrest and, in particular, to restore control of the forests. The summer of 1799, preceding the coup, was marked by a major drought.[59] Climate change even made the front page of one of the official newspapers. *Le Moniteur universel* published two doom-mongering articles signed by Cadet de Vaux, a famous chemist of the time. Cadet was alarmed by the disappearance of springs and the shrinking of rivers: 'We are devoured by drought and science says: we should blame not nature but Man who, in altering the surface of the Earth, has changed the course of the atmosphere and consequently the influence of the seasons.'[60] At this point too Rougier, who had become prefect of Yonne thanks to the Consulate, gave a resounding speech in Auxerre.[61] He warned against the *irreversible* character of climate change: 'Without trees we have no water and without water we have no trees.' 'There is a point', he added, 'at which nature rejects any recovery.'[62]

Obviously, exaggerating the threat also served to glorify the one individual capable of averting it. Thus, without naming him, Rougier called upon Napoleon, the 'hero' and 'philosopher' who in Egypt had witnessed first-hand the fate of nature when subjected to bad government. For the

58 Rougier de La Bergerie, *Traité d'agriculture pratique*, Paris, 1795, p. viii.

59 The meteorologist Louis Cotte showed that what was probably the driest summer of the century was coming to an end. Cf. Cotte, 'Notes sur la chaleur et la sécheresse extraordinaires de l'été de l'an VIII (1800)', *Journal de physique*, 1800, vol. 51, Messidor Year VIII, p. 216.

60 Cadet de Vaux, 'Observation sur la sécheresse actuelle, ses causes et les moyens de prévenir la progression de ce fléau', *Le Moniteur universel*, 1er fructidor an VIII, August 1800.

61 La Bergerie, *Mémoire et observations sur l'abus des défrichements*. A speech that was immediately reprinted by *Décade philosophique*, no. 18, year X, p. 513.

62 La Bergerie, *Mémoire et observations sur l'abus des défrichements*, p. 58.

glory of arms must give way to that of regeneration.[63] These interventions paved the way for the forestry reforms of the Consulate – in particular, the law of 9 Floréal Year XI (29 April 1803), which (re)imposed administrative authorization for land clearance and partially cancelled the deregulation of private woods decided in late 1791. The climate was one of the arguments justifying the end of the liberal parenthesis. In April 1803, Baron Delpierre declared in the Tribunate that the Revolution had been 'the occasion for senseless extirpation'; that it had caused a disastrous change of climate, even affected the health of our planet, for 'woods are an essential organ in the formation of the Earth'.[64]

Among the Cassandras of the Consulate, one author stands out for the quality of his intervention. This was Eusèbe de Salverte. Hailing from a family of financiers, he would subsequently pursue an important literary and political career. Prompted by the summer drought of 1799, he published *Conjectures sur la baisse apparente des eaux sur notre globe*.[65] His starting-point was the same as Cadet de Vaux's – namely, that rivers were shrinking – but he pushed the argument further. Water could not disappear: if it seemed to be becoming scarcer, it was because its *circulation* had slowed down. Salverte made an economic comparison: just as a country was poorer when money circulated badly, so it was drier when water circulated less quickly. To explain this phenomenon, he attempted a simulation of the globe's atmosphere. In a spherical, rotating bottle filled with water, he placed some wax shavings representing clouds. The shavings, being lighter, gathered at the poles due to centrifugal force. According to Salverte, this explained the increase in icecaps that caused harsh winters and droughts alike in Europe: water was retained at the poles. In Salverte, the complexity of nature's economy was no longer a pretext for lyrical flights on the Creator's infinite wisdom. It was a mechanical problem which he sought to understand by way of a small-scale model of the Earth, in a mimetic experiment intended to represent the global atmosphere.

A standard idea in political philosophy is that representative democracies were formed by externalizing environmental issues, declaring the issue of nature out of bounds, the better to focus on regulating human

63 Ibid., p. 64.

64 AN AD/IV, *Discours de Delpierre, tribunat, séance du 9 floréal an XI*, p. 4.

65 Eusèbe de Salverte, *Conjectures sur la baisse apparente des eaux sur notre globe*, Paris: Demonville, 1799.

relations. The environmental crisis would thus also be a crisis of political representation. Gutted of 'non-humans', modern representation is said to have rendered us blind to their destruction. This hypothesis hardly stands up to historical scrutiny: in France, it was precisely when the representative system was constructed that the issue of climate became political.

Indeed, it was the French Revolution that brought about the fundamental shift from colonial-era optimism to anxiety about environmental collapse. It was in revolutionary debates that climate warnings moved beyond naturalist circles to become a structuring element of the discourse in France on property, nature, and the state's role in their management. The climate alert was the cornerstone of a discourse on the natural order, aiming to govern rural populations in their relationship to environments after the end of feudalism. The climatic indictment of the ancien régime ultimately turned against the Revolution itself, inaugurating a half-century of debates over deforestation and climate change.

7

Climate Patriotism

The idea of a national climate became established in the late eighteenth century: people no longer referred exclusively to the climate of a town or province, or the climates of a kingdom, but to 'the climate of France', or Great Britain, or Spain, or even the United States of America. This was a general phenomenon directly linked to the construction of national imaginaries. It operated comparatively. The geographer Conrad Malte-Brun noted with irony that 'the various peoples of Europe took pride in attributing a material superiority to themselves as regards climate. . . . The Frenchman speaks with mocking pity of the unfortunate beer drinkers in Germany, and never fails to stress the moral effect of English fogs.'[1] Climate conferred a 'natural' homogeneity on composite political spaces or states that were in the process of being formed. For example, Juan Francisco de Masdeu, a Catalan Jesuit born in Palermo, opened his history of Spain (written in Italian) with a chapter 'on the climate of Spain and the Spanish genius for industry and literature'.[2] In South and Central American countries, the climate also helped produce national 'imagined communities' following independence.[3] If climate assisted in

1 Conrad Malte-Brun, *Précis de géographie universelle*, Paris: Aimé André, 1826, vol. 6, p. 42.

2 Juan Francisco de Madeu, *Historia crítica de Espãna y de la cultura española*, Madrid: Antonio Sancha, vol. 1, 1783, p. 1.

3 Benedict Anderson, *Imagined Communities: Reflections on the Origin and Spread of Nationalism*, London: Verso, 1982.

the construction of national identities, this was because it tied territory, people and history inseparably together.

The Climate of Independence

In the United States, climate patriotism promoted both American nature and the ability to create an improved nature, and hence humanity, through labour. This argument, which drew on imperial British theology, acquired new political importance with the prospect of independence. The latter was imagined in scholarly circles and salons before being won on the battlefield – for example, in the American Philosophical Society, founded by Benjamin Franklin in Philadelphia in 1747, which was revived in 1767 and counted among its members George Washington, John Adams, Thomas Jefferson and James Madison. American intellectuals set about producing a narrative that exalted the grandeur of nature in their continent and the civilizing activity of the pioneers.

This discourse was in part a riposte to the attack launched by Buffon on the degenerate nature of the New World.[4] According to the French naturalist, America was a colder and wetter continent than Europe, and consequently less conducive to the flourishing of higher life forms.[5] Animals there were smaller and weaker, and livestock brought over from Europe degenerated. This assessment extended to the human beings – Native Americans in the first instance, but also the colonists, who were doomed to physical and moral decrepitude. These ideas were taken up and amplified by European authors such as Cornelius de Pauw and William Robertson, who dwelt on the inferiority of America and Americans in famous, controversial books.[6]

This was the context in which, in August 1770, Hugh Williamson, a doctor from Pennsylvania, presented a paper on climate change to the

4 Antonello Gerbi, *The Dispute of the New World: The History of a Polemic, 1750–1900* (1955), Pittsburgh: Pittsburgh University Press, 2010.

5 Jacques Roger, 'Buffon, Jefferson et l'homme américain', *Bulletins et Mémoires de la Société d'anthropologie de Paris*, new series, vol. 1, instalments 3–4, 1989, pp. 57–65.

6 Cornelius de Pauw, *Recherches philosophiques sur les Américains*, 2 vols, Berlin: George Jacques Decker, 1768–9; William Robertson, *History of America*, 3 vols, Dublin: Whitestone, Watson, etc., 1777. Robertson was more nuanced than Pauw. For him the climate was more of a retarder or accelerator of societal progress than an absolute determinant. See Golinski, *British Weather and the Climate of Enlightenment*, pp. 179–81.

American Philosophical Society that made an impression. Williamson was close to Jefferson and would himself be a prominent figure in the War of Independence and one of the signatories of the US Constitution. In broad outline, his theory boiled down to claiming that deforestation equalized the temperature of different lands and seas and contributed to the abatement of winds. It was the same process that had warmed Italy's climate since antiquity, when the countryside was better cultivated. The explanation was to be found some distance away, in Germany, where in the meantime improvement in crops had been immense, cancelling out the disparity that had caused violent winds and cold in Rome.[7] Williamson skilfully utilized the idea of European climate change, promoted by Buffon, the better to refute his thesis of an 'inferior' American nature. The old continent had likewise endured a difficult climate before centuries of cultivation tempered its excesses. This was the very path the Whites of America proposed to take, and in just a few decades.

After US independence, climate change became a patriotic topic among American elites. Numerous authors took up their pens on the subject. The most illustrious of them was a future president: Thomas Jefferson. He was still only the governor of Virginia when, in 1781, the French consul requested a paper on his state's natural riches and commercial opportunities. Jefferson took advantage to compose a vibrant eulogy of American nature.[8]

His goal was to show that plants, animals, and men all prospered in America: rather than degenerating, nature became stronger. His estate at Monticello – 200 slaves working on 2,000 hectares of tobacco, wheat, and pastureland – provided him with a genuine testing ground. Jefferson, who kept a meteorological diary, noted a palpable change in the climate, with cold and hot spells alike becoming more moderate.[9]

---

7 Hugh Williamson, 'An Attempt to Account for the Change of Climate, which has been Observed in the Middle Colonies in North-America', *Transactions of the American Philosophical Society*, vol. 1, 1771, pp. 272–80. This text had a significant impact. It was reproduced in *Observations sur la physique et l'histoire naturelle*, vol. 1, 1773, pp. 430–36 and in *Journal des Sçavans*, October 1773, pp. 177–96.

8 Thomas Jefferson, *Notes on the State of Virginia*, London: John Stockdale, 1787. Golinski, *British Weather and the Climate of Enlightenment*, p. 200; James R. Fleming, *Historical Perspectives on Climate Change*, New York and Oxford: Oxford University Press, 1998, pp. 24–7; Roger, 'Buffon, Jefferson et l'homme américain'.

9 Jefferson, *Notes on the State of Virginia*, p. 134. Jefferson remained an assiduous weather watcher, encouraging his correspondents to do likewise and document the changing American climate. Cf. his letter to Lewis E. Beck, 16 July 1824, in *The Writings*

The virtues of the North American climate were also perceptible in nature. Jefferson favourably compared the size of the animals on his estate to European livestock; he praised the intelligence of Native Americans in light of that of the barbarians prior to the Roman conquest; he also equated the number of geniuses born on US soil, factoring in its three million inhabitants and short history, with those hatched by France and England over several centuries. From the weight of its pigs to the genius of Benjamin Franklin, the entire natural history of the USA attested to the beneficence of its climate. Jefferson even noted, with astonishment, that some of his slaves grew whiter over time. Under his pen, the United States of America became a Land of Cockaygne: 'a country of health and joy, peace and abundance, the soil is excellent, the climate salutary and agreeable, and the winters moderate and short. No country in the world is capable of greater improvement.'[10]

Many other authors championed the US climate, such as Samuel Williams, professor of mathematics and natural philosophy at Harvard,[11] or the doctors Benjamin Rush – combatant in the War of Independence and co-signatory of the Constitution – and William Curry. According to Curry, the United States was the only nation where the original dignity of the human species was on the point of being restored, thanks to liberty and climate improvement.[12] The French geographer and writer Constantin-François Volney, who travelled through the United States from 1795 to 1798, found public unanimity on the subject: 'both on the Atlantic coast and in the West, I collected the same witness statements, in Ohio, in Gallipolis, in Washington, in Lexington, in Cincinnati, in Louisville, in Niagara, in Albany, everyone repeated the same circumstances to me – longer summers, later autumns, shorter winters . . .. And these changes were depicted to me not as gradual and progressive, but as rapid and almost sudden, commensurate with the extent of deforestation.'[13]

---

of Thomas Jefferson, Washington: The Thomas Jefferson Memorial Association, 1903, vol. 16, pp. 71–4.

10 Cited in d'Arcy Wood, *L'Année sans été*, p. 212.

11 Samuel Williams, *Natural and Civil History of Vermont* [1794], Burlington: Mills, 1809, vol. 1, p. 73. According to his meteorological diary, Williams estimated that the temperature had risen by 6° in Cambridge (Massachusetts) in the 150 years since the arrival of the colonists. Cf. Zilberstein, *A Temperate Empire*, pp. 164–8.

12 William Curry, *An Historical Account of the Climates and Diseases of the United States of America*, Philadelphia: T. Dobson, 1792, pp. 79–93, 398–407.

13 Constantin-François Volney, *Tableau du sol et du climat des États-Unis*, Paris, 1803, vol. 1, p. 290.

In 1811, Hugh Williamson returned to the question he had posed forty years earlier. After a long career in Congress, he once again wanted to refute rumours of degeneration, for this (he explained) was to do 'his duty as a citizen of the United States', as a 'patriot and father'.[14] In short, he was to write a *patriotic climatology*.

In the United States, the discourse on climate change was clearly part of a process of nation-building that exalted both a population – white, Protestant, industrious – and the nature that population moulded and was strengthened by in return. There were two aspects to this ideology: it celebrated self-fashioning (of individuals, of a people) and a Promethean mastery of nature. Improved by independence, the American climate was the best in the world and the United States was the cradle of a new race of men, strong and free, on a par with the ancient Greeks. Something fundamental was at stake in this exaltation of human action. The US political community presented itself as the product not so much of natural determinants as of a twofold dynamic of self-fashioning: of itself and of its material environment. The United States was not only a developing neo-Europe, but the site of all possible ameliorations.

The Climate of 'Improvement'

In Great Britain, the question of climate change did not arouse the same passions, patriotic or declinist, as it did in the United States or France. This was not because the weather was more clement there. The British, too, suffered a Little Ice Age: in 1782, 1784, 1788, and once again in 1798–9, harsh winters and/or wet summers led to price increases and food riots. Nor was it because British scientists were uninterested in the issue. On the contrary, the water cycle was fully integrated into natural theology and was a kind of commonplace among eighteenth-century British scientists. For example, Gilbert White, author of the veritable bestseller *The Natural History of Selborne* (1784), whose success endures to the present day, made a classical comparison: 'In heavy fogs, on elevated situations, trees are perfect alembics.'[15]

14 Hugh Williamson, *Observations on the Climate in Different Parts of America*, New York: T. & J. Swords, 1811, pp. iii, 2.

15 Gilbert White, *The Natural History of Selborne* [1784], Philadelphia: Carey & Lea, 1832, p. 242.

The problem was, rather, that climatic discourse gained virtually no political purchase. First of all, deforestation, older and much more advanced than in France, was not regarded by the English as a brutal rupture of the natural order. The prospect of a shortage of wood was tempered by coal and massive imports from the Baltic (and then, during the continental blockade, from Canada). Next, the British gentry was attached to an ideology of 'improvement' which justified agrarian innovations – enclosures, drainage of marshes and artificial pasture – and its own economic privileges. Thus an author explained that soils and climates improved at the same time thanks to 'an assiduous attention to their improvement'.[16] In 1818, the agronomist and statistician John Sinclair asserted that deforestation *tempered* climates, whereas in France it was thought to worsen them.[17] Finally, the climatic accidents of the late eighteenth century – freezing winters or unduly wet summers – were attributed to external causes, whether the explosion of the Icelandic volcano Laki (1783) or the appearance of icebergs in the North Atlantic.[18] And, even in the latter instance, the idea of improvement persisted, as in the proposal of Erasmus Darwin (grandfather of Charles) to employ Europe's fleets to tow the icebergs to lower latitudes. Thanks to this naval geo-engineering, European winters would be made milder, while tropical countries would enjoy a welcome freshness![19]

The possibility of climatic deterioration concerned the Irish and Scottish peripheries above all. As regards Ireland, Reverend Hamilton of Dublin worried about the growing violence of the winds due to coastal deforestation – fears rapidly dispelled by contemplating the 'athletic frame' and 'ardent passions' of his compatriots, which (it seems) were 'celebrated throughout Europe'.[20] Similarly in Scotland, in the 1790s, several priests reported climate cooling. No complaint was more widespread in the Highlands, claimed one of them, who advanced serious evidence: repeated flooding, gullied riverbeds, and the later arrival of migrating birds. The

16 John Campbell, *A Political Survey of Britain*, London, 1774, p. 60.

17 John Sinclair, *Code of Agriculture*, Hartford: Hudson, 1818, p. 4.

18 One of the first to venture this hypothesis, which was very popular in the eighteenth and nineteenth centuries, was Richard Bradley, professor of botany at Cambridge and a member of the Royal Society. Cf. Bradley, *A General Treatise of Husbandry and Gardening*, London: Woodward, 1726, vol. 2, p. 437.

19 Erasmus Darwin, *The Botanic Garden*, New York: Sword, 1798, p. 32.

20 William Hamilton, 'Memoir on the Climate of Ireland', *Journal of Natural Philosophy, Chemistry, and the Arts*, vol. 2, December 1798, p. 435.

deforestation of Scottish mountains was implicated. But these complaints sent to the Board of Agriculture were not followed up. Tasked with dealing with them, John Sinclair doubted these allegations, which in any event did not concern the agricultural lowlands.[21]

If the English elites were content with their climate, they still did not make it a central element of patriotic discourse. The English climate was perceived as salubrious (apart, obviously, from London, shrouded in smog by the burning of coal) and suited to agriculture, but nevertheless inferior to that of France. Thus, the agronomist Arthur Young explained that he preferred the latter to England's, for in it vines, olive trees and maize could be grown. But this French superiority was only apparent: the proverbial variability of the English climate was held to forge a superior national character, at once hardy and tenacious, and, crucially, sufficiently *adaptable* to be capable of colonizing territories in every corner of the world. India afforded a perfect contrast: its emollient tropical climate prepared its inhabitants for subjugation.[22] The idea of British climatic particularism also took root: the insular climate, changeable but fortifying, was supposedly characteristic of the United Kingdom, from Dublin to London to Edinburgh. Stressing climatic homogeneity, this discourse also played a part in fabricating a certain national unity after the Act of Union between England and Scotland in 1707.

It was this backdrop of self-satisfaction that one John Williams (1773–1853) sought to shatter at the start of the nineteenth century. This horticulturist from Worcester, who prided himself on his knowledge of science and agricultural innovations, is virtually unknown to historians. Yet his *Climate of Great Britain*, published in 1806, was one of the most ambitious works of the age on climate.[23]

Williams formed part of a small elite of scientific gentleman farmers. He was a member of the Royal Horticultural Society (founded in 1804) and very close to its president, Thomas Andrew Knight, an important figure in the improvement of agricultural varieties.[24] For Williams the climate

21 Jonsson, *Enlightenment's Frontier*, pp. 77–81.

22 Golinski, *British Weather and the Climate of Enlightenment*, pp. 52–66.

23 John Williams, *The Climate of Great Britain, or, Remarks on the Change It Has Undergone, Particularly in the Last Fifty Years*, London: Baldwin, 1806. Jankovic mentions his plan for dehumidifying England in *Reading the Skies*, p. 1.

24 In chapter 1 of *The Origin of Species*, Charles Darwin cited Knight as a pioneer of the theory of evolution; his experiments on heredity anticipated those of Mendel. Cf. Murray Mylechreest, 'Thomas Andrew Knight (1759–1838) and the Application of

issue was therefore primarily practical: Worcestershire was a major centre of fruit production, and orchards are much more sensitive to meteorological circumstances than other crops. How, then, were spring frosts to be forecast? Should orchards be heated? Or ripening accelerated in order to sell the fruit as early seasonal produce?[25] But the book's argument went far beyond horticulture. It formed part of a series of debates on the impact of enclosures and the Corn Laws. A paradox haunted John Williams: how was it that, despite all the agronomic improvements, the price of cereals was increasing and the English people were going hungry? Unlike his celebrated contemporary Thomas Malthus, the explanation he advanced was not demographic but climatic: over the last thirty years, the English climate had become too damp.

How was this deterioration to be explained? The horticultural greenhouses of Worcester – his neighbour owned more than a hectare of plantations under glass[26] – provided him with an analogy: Great Britain was like an 'immense hothouse' that had been poorly laid out, an 'over-crowded hothouse' in which 'a great variety of vegetables [are] confined within a limited atmosphere'.[27]

The first cause of the growing humidity was the introduction of exotic plants from Italy and the Levant. Elm and hawthorn, in particular, released excessive quantities of water into the atmosphere, given the already damp climate of England. Botanical globalization had disrupted the divine arrangement of plants, which was intended to temper climates.[28] A second cause was enclosures, with the proliferation of grazing fields and their hedges crisscrossing the land. By dint of this transformation of property regimes, England was suffering from an *over-abundance* of flora. Williams backed up his claim by linking the number of enclosure laws since the eighteenth century to signs of climate cooling.[29] Third cause: the 'soil-less' nature of the British economy. Starting from the deficit in the trade

Experimentation to Horticulture', *Annals of the History and Philosophy of Biology*, vol. 15, 2010, pp. 15–27.

25 James Main, *Horticultural Register*, vol. 4, 1835, p. 203; John Williams, 'On the Method of Hastening the Maturation of Grapes. Letter from John Williams to Joseph Banks', *Transactions of the Horticultural Society*, vol. 1, 1812, pp. 107–12.

26 Bill Gwilliam, *Old Worcester: People and Places*, Broosgrove: Halfshire Books, 1993, p. 189.

27 Williams, 'On the Method of Hastening the Maturation', pp. 108–11.

28 Ibid., p. 30.

29 Ibid., chapter 12.

balance, Williams explained that England imported a growing mass of vegetable products (grains, sugar, wine, etc.), which, ingested, digested, or transformed, fertilized the soil to excess. They nourished an exuberant flora, saturating the atmosphere with vapour.

John Williams's book was received with indifference, even condescension. Reviews overlooked his critique of enclosures and free trade, instead mocking his plan – very striking, it is true – to dot the country with mills which, activating woollen pads on glass cylinders, would electrify the atmosphere and dry out the climate. Whereas Williams sought to link his work to cereal prices, the food riots of 1804, and the debate on the Corn Laws, agrarian economists – James Maitland de Lauderdale and William Spencer – ignored his climatic argument in favour of protectionism.[30] A subsequent episode further attests to the reticence of English scholars about his theory. In 1818, after the 'summerless year' caused by the Tambora volcano (see Chapter 8), John Williams again tried to promote his ideas on climate change. He hoped to present a paper to the Royal Society and put out feelers to his friend Thomas Knight, member of that prestigious institution. Knight's response is revealing. While he was willing to pass the text on to Joseph Banks, the powerful president of the Society, he warned Williams about the heterodox character of his ideas: 'The fact that Europe has grown milder by the destruction of its forests appears to be universally accepted.'[31] Williams's climate theory so contradicted the British elites' ideology of improvement that it was doomed to failure.

Across the Channel, people congratulated themselves just as much on their fine climate. In the late eighteenth century, French elites regarded their climate as the best in Europe, at once temperate, varied, well-watered and healthy, making France a great agricultural country capable of feeding itself. The concise article on 'France' in Diderot and d'Alembert's *Encyclopédie* explained that 'the air there is pure and healthy, under a sky that virtually everywhere is temperate . . . Its fertile, delightful countryside abounds in salt, grains, vegetables, fruits, wines, etc.'. This indulgent judgement was picked up on by agronomic elites, who stressed

30 Ronald Meek, *The Economics of Physiocracy*, Cambridge, MA: Harvard University Press, 1963, pp. 323–414.

31 Letter of 21 January 1818 from Thomas Knight to John Williams, in Thomas Andrew Knight, *A Selection from the Physiological and Horticultural Papers*, London: Longman, 1841, pp. 37–8.

the discrepancy between the climate and the shortages suffered by the people. Something was not right, and they intended to resolve it with their agronomic reforms and liberalization of the cereal trade.

At the beginning of the following century, the major theme of French climatic patriotism was variety. In 1826 the geographer Conrad Malte-Brun introduced the terms (the idea is older) of 'oceanic climate', 'continental climate' and 'Mediterranean climate'.[32] In Europe, France was said to be the blessed zone where these different influences met. This climatic variety, which also meant a variety of plants and terroirs, became a kind of national particularity vaunted by historians and geographers – Jules Michelet, Paul Vidal de La Blache – and taught to generations of schoolchildren. So varied a climate was also held to have moulded a people that was more sensitive, more lively, quicker to be moved artistically and politically than its neighbours to the North or the East, who were serious and hard-working but dull. Finally, this wonderful climate was a recapitulation of history: the reward for the toil of a peasant people. In nineteenth-century France, it was commonplace to insist on the ages-old working of the soil and land that had supposedly made the climate milder. The diagnosis of a deterioration in the climate caused a trauma in France, as intense as the emotions the elites had invested in it.

32 Malte-Brun, *Précis de géographie universelle*.

8

In the Shadow of the Volcano

On the evening of 10 April 1815, the Tambora Volcano started to erupt on the island of Sumbawa, situated to the east of Java, in present-day Indonesia. The explosion shot a plume of ashes, debris and sulphur into the atmosphere. It was one of the biggest volcanic eruptions in human history. The billions of tons of matter expelled disrupted the planet's atmospheric system for years to come.[1] While the explosion of Tambora created the magnificent orangey-red skies painted by William Turner at the time, a veil of blood and death was cast over the whole globe. So cold and wet was the season that, in many countries, 1816 has gone down in history as 'the year without a summer'.[2] Yet this phrase, more evocative of a rainy August than a climatic and demographic crisis, minimizes to a stupendous degree the suffering, gnawing hunger, and political turbulence of that period.

A year after the rotten summer of 1816, during the 'lean season',[3] a journalist noted that 'never has the civil existence of nations seemed

1 Clive Oppenheimer, 'Climatic, Environmental and Human Consequences of the Largest Known Historic Eruption: Tambora Volcano (Indonesia) 1815', *Progress in Physical Geography*, vol. 27, no. 2, 2003, pp. 230–59; Le Roy Ladurie, *Histoire humaine et comparée du climat*, vol. 2, 'Disette et Révolution' (1740–1860)', pp. 277–308.

2 C.R. Harington, ed., *The Year without a Summer? World Climate in 1816*, Ottawa: Canadian Museum of Nature, 1992. Climatologists assessed the reduction in global temperature due to the explosion to be 0.7° C.

3 The agricultural lean season is the period preceding the first harvests of the year, when grain from the previous harvest might be in short supply.

more bound up with their physical existence'.[4] Food supply problems throughout Europe linked meteorology closely to the political order. In the revolutionary cold climate of 1816–17, the European scientific elite systematically emphasized the *stability* of the climate over and above the concrete experience of cooling. For the established powers, it was crucial to underplay the phenomenon and demonstrate its transient character. For if natural structures should change, the whole political order could be called into question.

A Planetary Catastrophe

The explosion of Tambora was, above all, a terrible catastrophe that devastated the island of Sumbawa. It wiped the principalities of Tambora, Pekat and Sangaar off the map.[5] The only first-hand testimony that has come down to us is from the prince of Sangaar:

> Three distinct columns of flame burst forth . . . In a short time the whole Mountain next to Saugur appeared like a body of liquid fire extending itself in every direction. Stones . . . fell very thick, some of them as large as two fists . . . between 9 and 10 p.m. ashes began to fall and soon after a violent whirlwind ensued which blew down nearly every house . . . tearing up by the roots the largest trees, and carrying them into the air together with men, houses and cattle.[6]

British garrisons posted thousands of kilometres away thought they heard a canon going off. A rain of ashes destroyed harvests and the survivors had to face the agonies of hunger. 'Entire villages were abandoned as the survivors dispersed in search of food', reported a British naval officer. Khatib Lukman, a Malayan man of letters officiating at the royal court of Bima (the island's principal town), reported how Javan, Arab, Chinese and Dutch merchants seized the treasures of Sumbawa in exchange for

4 *Gazette de Lausanne*, 8 July 1817.

5 For a general history of the effects of 1816, see d'Arcy Wood, *L'Année sans été* and Bernice de Jong Boers, 'Mount Tambora Explosion in 1815: A Volcanic Eruption in Indonesia and Its Aftermath', *Indonesia*, no. 60, 1995, pp. 37–60.

6 J.T. Ross, 'Narrative of the Effects of the Eruption from the Tomboro [*sic*] Mountain in the Island of Sumbawa', *Batavian Transactions*, no. 8, 1816, pp. 1–25.

foodstuffs: 'All these lands enriched themselves by buying arms, ceremonial clothes and slaves by the hundreds; the Bimanese were powerless.' Giving oneself up to slavery was often the only way to survive.[7] According to the calculations of a Swiss scientist who visited Sumbawa thirty years after the explosion, the island had lost half of its inhabitants: 10,000 in the eruption, 38,000 as a result of famine, and 36,000 through exile.[8]

As regards the global climate, the debris and ashes scarcely mattered because they fell back to earth after a few days, washed by the rains. However, the sulphur aerosols injected into the atmosphere travelled the whole globe, leaving traces in sediments and even in the ice at the two poles. Combining with water molecules, the sulphur created a mist, reflecting a percentage of solar energy towards space and lowering world temperatures.

In 1815, the mist from Tambora settled over a South China already weakened by the previous year's drought. Granary reserves were already low when intense rains in late spring 1815, followed by an August frost, destroyed two successive harvests. Ordinarily, the peasants of the region obtained very high yields thanks to biennial, or even triennial, rice harvests. But their reliance on a single cereal also rendered them highly vulnerable to the vagaries of the weather. The ensuing demographic catastrophe, difficult to quantify though it is, is now cited by historians to explain the economic divergences between China and Western Europe that widened in the nineteenth century.[9]

In India, the 1816 monsoon was deficient. The north of the country suffered from drought, as well as abnormally low temperatures.[10] By contrast, the following year was marked by devastating floods. During the summer of 1817, thousands of villages along the Ganges and in its delta were invaded by the waters. It was also in 1817 that India experienced its first major cholera epidemic, a disease hitherto confined to the Bay of

---

7 Hubert Chambert-Loir, 'Mythes et archives. L'historiographie indonésienne vue de Bima', *Bulletin de l'École française d'Extrême-Orient*, vol. 87, no. 1, 2000, pp. 212–45 (especially p. 236).

8 Heinrich Zollinger, *Besteigung des Vulkanes Tambora auf der Insel Sumbawa und Schildung der Erupzion desselben im Jahr 1815*, Winterthur: Wurster, 1855, p. 20.

9 D'Arcy Wood, *L'Année sans été*, pp. 115–43 and Shuji Cao, Yushang Li and Bin Yang, 'Mt Tambora, Climatic Changes, and China's Decline in the Nineteenth Century', *Journal of World History*, vol. 23, no. 3, 2012, pp. 587–607.

10 *The Asiatic Journal*, vol. 4, 1831, p. 166. On the link between Tambora and cholera, see once again d'Arcy Wood, *L'Année sans été*, pp. 91–117.

Bengal. For the European doctors of the time, trained in neo-Hippocratic medicine, climatic imbalance was the obvious cause of the cholera that spread through the world in subsequent decades.[11]

In Europe, the years 1816 and 1817 saw the last major subsistence crisis on a continental scale.[12] Nevertheless, populations were very differently affected. In England, for example, the import of wheat from Ireland, which was itself prey to famine, cushioned the crisis. The price of the cereal doubled in Switzerland, in some parts of Prussia, and in the Austro-Hungarian Empire. But it only rose by 46 per cent in Great Britain.

Annual national averages conceal more significant local disparities and monthly variations. At the nadir of the crisis in June 1817, the price of wheat multiplied by 2.5 in France and 4.5 in Switzerland. Such a hike was a calamity for the popular classes who, *in normal times*, spent half their income on bread. Inequalities grew: big farmers profited from the increase, while day labourers and urban workers saw their purchasing power collapse.

Up to the autumn of 1817, markets were the scene of food riots throughout Western Europe. These riots were generally part and parcel of a 'moral economy': the population forced traders to hand over their goods at a price it reckoned 'fair'. Thus, in June 1817, at least 4,000 people overran Château-Thierry, a key market supplying Paris. 'The grain merchants are pillaged with unimaginable orderliness', wrote the sub-prefect.[13] The mayor was sometimes called on to organize 'popular taxation' of traders or big farmers. The riots were primarily aimed at preventing grain leaving for other towns – Paris in particular. This explains why numerous troubles occurred in agricultural regions when the risk of shortages was remote.[14] Even in England, comparatively sheltered from a dearth of food, rioters looted mills and blocked the departure of cereals for London.

---

11 James Jameson, *Report on the Epidemik Cholera Morbus*, Calcutta: Government Press, 1820, p. LXVIII. Contemporary biomedical research is re-examining the climatic causes of major nineteenth-century epidemics by studying the transformation in the ecology of the vibrio of cholera and its genetic mutations. Cf. Rita Colwell, 'Global Climate and Infectious Disease: The Cholera Paradigm', *Science*, vol. 274, 1996, pp. 2025–31.

12 John D. Post, *The Last Great Subsistence Crisis in the Western World*, Baltimore and London: Johns Hopkins University Press, 1977.

13 Quoted in Robert Marjolin, 'Troubles suscités par la disette de 1816–1817', *Revue d'histoire moderne*, vol. 8, no. 10, 1933, p. 438.

14 Nicolas Bourguignat, 'Le maire nourricier: renouvellements et déclin d'une figure tutélaire dans la France du XIXe siècle', *Le Mouvement social*, no. 224, 2008, pp. 89–104; Louis Gueneau, 'La disette de 1816–1817 dans une région productive de blé, la Brie', *Revue d'histoire moderne*, vol. 4, no. 19, 1929, pp. 18–46.

A Providential Debacle

It was against this background of serious disorder that the idea of climate cooling won over much of public opinion in England, Switzerland and France.

In England, where the newspapers generally tended to minimize agricultural problems for fear of fuelling grain speculation, reports of bad harvests nevertheless filtered into the press. In October 1817, the *Morning Chronicle* published a long article on the 'deterioration of the climate', linking the phenomenon to the planetary cooling foreseen by Buffon. The multiplication of icebergs on trans-Atlantic routes was said to prove the reality of the process.[15]

However, just as climate cooling was becoming palpable, London's scientific elite was concerned above all to announce signs of *warming*. Proof was to be found in the Arctic ice. In late summer 1817, the crews of whaling ships reported sensational news: the ice shelf in the north of the island of Spitzberg, where they stayed during the fishing season, had entirely disappeared. William Scoresby, one of the great whalers of the era, explained that at 72 degrees north latitude, rather than encountering ice and whales as usual, he found the ocean completely clear for 200 square leagues. The news was reassuring: while revolt was rumbling and bread riots proliferating, climate development seemed to be reversing and nature was finally resuming its normal course.

Joseph Banks, president of the Royal Society, asked Scoresby for further details. This debacle, he explained, augured well for the English climate, for it seemed clear to him that the 'frosty springs and chilly summers we have been subject to for many years are caused by the increase of ice which seems to have accumulated for many years past'.[16] For Banks, the issue was all the more urgent in that he was a spokesman for large landowners. He defended agrarian protectionism (the Corn Law of 1815) and promoted acclimatization experiments aimed at making England self-sufficient.

15 On 20 July 1816, *The Times* briefly noted that if the present rainy weather were to persist, the consequences were bound to be ruinous for people in general. Other papers preferred to blame the protectionism of continental countries for the economic slump. See Post, *The Last Great Subsistence Crisis*, pp. 16–28; *Farmer's Magazine*, 1818, p. 168.

16 Joseph Banks to William Scoresby, 22 September 1817, in *The Letters of Sir Joseph Banks: A Selection, 1760–1820*, ed. Neil Chambers, London: Imperial College Press, 2000, p. 329.

This political line would become highly problematic if the English climate continued to cool.[17]

Significantly, most articles on the English climate appeared *after* autumn 1817 – that is, *after* the crisis – so as to convey the good news from the Arctic. The return to climatic order was hailed by Thomas Brande, an intimate of Banks and a professor at London's Royal Institution. Founded by City merchants in 1805, the Royal Institution was flourishing, popularizing in Mayfair, at the heart of the English capital, a science that aspired to be both practical and spectacular. Brande and Banks were also members of the Board of Agriculture, a para-governmental institution responsible for defending agricultural protectionism. Having recalled the 'terrible ordeals' of climate cooling over 400 years (icebergs, the growth of glaciers, and the disappearance of vines from England), Brande dramatized the return to normality represented, in his view, by the polar debacle of 1817.[18] His article enjoyed quite extraordinary success. It was reprinted *in extenso* (often under the more accurate title of 'Alleged Deterioration of Climate') by the *Gentleman's Magazine*,[19] the *Farmer's Magazine*,[20] and the *Literary Panorama*,[21] as well as in the whole local press.[22] And, in the wake of Brande, all insisted on the impossibility of a general global cooling.[23] The *Chester Chronicle* asserted in peremptory fashion that 'nothing could be effecting such a change, except some power that was constantly producing a still greater inclination of the axis of the earth'. The *Farmer's Magazine* explained that the climate was a system in constant motion yet perfectly stable. The planet was stable, because it was enormous and very old: 'That

17 John Gascoigne, *Science in the Service of Empire: Joseph Banks, the British State and the Uses of Science in the Age of Revolution*, Cambridge: Cambridge University Press, 1998, pp. 115–17.

18 Thomas Brande, 'Some Remarks on the Deterioration of the Climate of Britain, with an Attempt to Point Out its Cause', *The Journal of Science and the Arts*, vol. 4, 1818, pp. 281–7.

19 *Gentleman's Magazine*, vol. 88, 1818, pp. 134–6.

20 *Farmer's Magazine*, 1818, p. 72.

21 *Literary Panorama*, vol. 8, p. 89.

22 *Carlisle Patriot, Bath Chronicle, Cambridge Chronicle Journal, Stamford Mercury, Northampton Mercury, Hull Packet, Perthshire Courier, Durham Country Advertiser, Worcester Journal, Morning Advertiser, Leeds Intelligencer, Cheltenham Chronicle, Taunton Courier, Kentish Weekly Post.*

23 See, for example, 'Refutation of the Alleged Deterioration of Climate', *Chester Chronicle*, 27 March 1818; 'Climate of Great Britain', *Cheltenham Chronicle*, 7 May 1818; *The Scots Magazine*, 1 July 1818; *Perthshire Courier*, 20 August 1818; *Edinburgh Magazine*, vol. 82, 1818, p. 71.

machinery, however vast and complex, must long since have acquired a permanent state of working'.[24]

In subsequent years, the question of global climate, its warming or its stability, continued to occupy the British public through the major issue of the North-West passage, a maritime route awaiting discovery that would enable European ships to reach Asia by skirting Northern Canada.[25] John Barrow, second secretary at the British Admiralty, was its great promoter. He defended his project by insisting on the end of climate cooling, a development that would make navigation in the Far North possible (if not easy). His interventions were controversial. His main opponent, the Scottish physician and mathematician John Leslie, hit back with a study in retrospective climatology that was extracted in *The Times* and, given the post-Tambora context, widely publicised in Europe. Leslie denied the existence of any climate change in Europe, on the basis of a reconstruction of 'harsh winters' that drew on works of historical climatology from 1770–1790 (Chapter 4).[26]

Reassuring Glaciers

Switzerland was one of the European countries worst hit by the consequences of the eruption of Tambora. Hunger in the mountains, shortages in the plains, and urban riots were commonplace in these years. Mortality increased by more than 50 per cent (only 2 per cent in France).[27] As a result of the economic segmentation of the country, certain eastern cantons were more severely affected: mortality doubled in Saint-Gall and Appenzell. A pastor described 'skeletal men devouring the most disgusting fare – bodies, nettles, food they fought animals for'.[28] In these years,

24 'Remarks on Climate', *Farmer's Magazine*, 1818, p. 324.

25 Siobhan Carroll, 'Crusades against Frost: Frankenstein, Polar Ice, and Climate Change in 1818', *European Romantic Review*, vol. 24, no. 2, 2013, pp. 211–30; Anya Zilberstein, 'A Considerable Change of Climate: Glacial Retreat and British Policy in the Early-Nineteenth-Century Arctic', in Miglietti and Morgan, eds, *Governing the Environment in the Early Modern World.*

26 John Leslie, 'Polar Ice, and a North-West Passage', *Edinburgh Review*, vol. 30, July 1818, pp. 1–59.

27 Post, *The Last Great Subsistence Crisis*, p. 114 and Le Roy Ladurie, *Histoire humaine et comparée du climat*, vol. 2, p. 303.

28 *Gazette de Lausanne*, 11 April 1817, p. 2.

Switzerland experienced a major wave of emigration to the United States and Russia.

Cooling posed special problems for the Alpine economy. With grazing land covered in snow in summer, many animals starved to death.[29] Some owners of flocks forced their shepherds to go up to the summer pastures despite the appalling weather. Rebellion rumbled in the mountains.[30] In the plains, too, cooling was a source of anxiety. Property-owners became alarmed about the future of agriculture and of grapevines in particular: 'Might there be some truth to the progressive cooling of the globe in Buffon's system?' wondered a Swiss agronomist.[31]

As elsewhere in Europe, different climatic theories were current and filtered into the press. Some blamed lightning rods,[32] others icebergs drifting southwards,[33] others still Buffon's cooling.[34] The seemingly most fashionable explanation pointed to sunspots. During the summer of 1816, these were so big that they could be made out with a simple, smoke-blackened glass slide.[35] The rumour mill was in full swing: the end of the world was nigh. One brochure announced it for 18 July 1816, affirming that on that day a piece of the Sun would crash into Earth. The *Gazette de Lausanne* mocked these predictions à la Nostradamus, while acknowledging that 'in truth we do not know what is happening in or on our globe: it is trembling everywhere and it freezes in summer; we are no longer conversant with it.'[36] In Switzerland, as elsewhere, newspapers strove to reassure the public. A short article on historical climatology ended on an optimistic note: 'The temperature that is causing Europe distress today is far from being unprecedented. All centuries have had their calamitous weather.'[37]

29 Ibid., 24 September 1816.

30 Marc Henriod, 'L'année de la misère en Suisse et plus particulièrement dans le canton de Vaud, 1816–1817', *Revue historique vaudoise*, 1917, p. 139.

31 Moïse Matthey-Doret, *Mémoire sur les moyens de prévenir la disette en Suisse*, Vevey: Loertscher, 1820, p. 23.

32 *Gazette de Lausanne*, 2 February 1816, p. 3.

33 Ibid., 31 July 1818.

34 Ibid., 25 June 1816.

35 Ibid., 17 September 1816.

36 Ibid., 28 August 1817.

37 Ibid., 2 August 1817. The *Gazette de Lausanne* devoted several articles to sunspots, recalling that the phenomenon had been observed by all astronomers since Galileo and appeared to have no effect on the terrestrial climate. Cf. ibid., 5, 16 and 23 July 1816, 30 August 1816.

Glaciers, in particular, furnished the naturalist elite with helpful arguments against gloom and doom. On 6 October 1817, the Helvetic Natural Science Society, which had just been founded in Geneva, decided that the cooling of the Alps would be the subject of its very first prize competition. According to its records, the subject imposed itself for two reasons: its importance 'for the general physics of the globe' and 'for the breeding of livestock'. The wording of the competition expressed a certain bias against Buffon. The methods of inquiry inherited from eighteenth-century climate debates were not forgotten: contributions were expected to address, (1) the lowering of the snowline; (2) variations in the stratification of vegetation, such as forests that had disappeared or grazing land that had become sterile; and (3) glacial moraines.[38] In short, the Helvetic Natural Science Society sought to revive the empirical glaciology of the late eighteenth century, in order to refute Buffon and reassure the Swiss people. What it did not anticipate was that its competition would spearhead one of the greatest advances in nineteenth-century geology: the theory of ice ages.[39]

In 1820, it only received two papers. The first, entitled *La Décadence de la nature*, was by the Genevan doctor Georges-Chrétien Desloges.[40] The author posited an original thesis. The centrifugal force produced by the Earth's rotation resulted in its gradual expansion. All locations were therefore gaining in altitude, the temperature was falling, and nature was degenerating. Whatever the merits of his explanation, the author's documentary efforts were remarkable: he itemized the villages where wheat cultivation had been abandoned and grazing land conquered by glaciers. He went so far as to consult the baptismal registers of Zermatt and Evolène – two villages today separated by impassable glaciers – to show that marriages between them were frequent in the fifteenth century. But the Helvetic Society naturally rejected this work ('a collection of scattered facts'),[41] which demonstrated the opposite of what it expected.

The paper that received the prize was very different. Its author, Karl Kasthofer, sought to prove the global stability of the climate. The popular thesis of cooling seemed sound at first sight: the dates of summer pasturage had been put back, a village famous for its kirsch no longer had cherry

38 Archives of the State of Berne, GA SNAW, 170, 7 October 1817.

39 Krüger, *Discovering the Ice Ages*, p. 98.

40 Georges-Chrétien Desloges, *La Décadence de la nature*, 1819, p. 16.

41 Archives of the State of Berne, GA SNAW, 170, 25 July 1820.

trees, another remained burdened with a mill tax when growing wheat was impossible, and so on.[42]

However, Kasthofer reckoned, caution was in order – and this for several reasons. The first was psychological: 'It is natural for man to forget past ills', and embellish the nature of yesteryear. From this derived 'a tradition that had spread to all the valleys, that of the *Blumlis Alp* [Alps in bloom], today covered with ice that once offered green pastures'. Second, the abandonment of a crop or the closure of a pass did not necessarily prove cooling, but might be linked to a change in 'commercial and industrial relations'. Third, climate cooling was certainly not ubiquitous: while the climate had perhaps cooled in the Swiss Alps, elsewhere it might have warmed. Finally, and most decisively, furrows in the lateral rocks overhanging glaciers, as well as moraines located downstream, indicated that glaciers had been far more massive in the past. According to Kasthofer, this proved that over 'millions of years' the temperature rose and fell periodically, but irregularly, with no possibility of prediction.

In fact, what primarily concerned Kasthofer, who was a forestry administrator, was the degradation of ecosystems by mountain communities. Rather than climate change being a cause of nature's decline, the reverse was true. The destruction of forests gave free rein to the wind and the cold, which progressively sterilized grazing grounds. Their disappearance therefore stemmed not from an external dynamic, but from their over-exploitation and the lack of forestry policing. The Helvetic Society subscribed to this explanation and stressed the negative role of the *shared* management of grazing lands. The disasters attributed to 'gradual climate change' were in fact due to deforestation: snow slid down slopes, created avalanches and swelled glaciers. The source of the evil was the spirit of independence among mountain folk, 'who, jealous of their liberty, refuse to accept into their midst the activity and surveillance of the police'.[43]

During the summer of 1818, one glacier in particular worried the Helvetic Society. The Giétroz, downstream of Val de Bagnes, had expanded so much since 1815 that it had cut off the bed of a river – the Dranse – thus forming an enormous water reservoir of 20 million cubic metres. In the event of a breach, the whole valley could be devastated. Ignace Venetz, an

42 Karl Kasthofer, 'Mémoire sur les changements du climat des Alpes', *Nouvelles Annales de voyages*, vol. 29, 1826, pp. 369–90.

43 Archives of the State of Berne, GA SANW, 170, 26 July 1820; Conrad Escher et al., *Rapport sur l'état actuel de la Vallée de Bagne*, Zurich, 1821, p. 61.

engineer from the canton of Valais, who was also a member of the Helvetic Society, was despatched to direct a highly delicate operation: emptying the lake by drilling a drainage tunnel in the glacier. Despite dogged work in perilous conditions, the glacier dyke gave way on 16 June. An enormous wave of water, ice and mud submerged Val de Bagnes, sweeping away everything in its path. The number of victims was put at forty-four.[44] This disaster, reported by all European newspapers, made the cooling of Switzerland a tangible issue.

In 1821, in turn, the engineer Venetz submitted a (belated) response to the competition of 1817.[45] He shared the Helvetic Society's aim ('to strongly contest the general opinion on the cooling of our Alps') and adopted for his own purposes Kasthofer's indictment of the mountain communities. Above all, he stressed a point that seemed to him fundamental: the considerable distances that separated present-day glaciers from their old moraines. 'In an age lost in the mists of time', he concluded, 'gigantic', 'inordinate' glaciers had covered the Swiss Alps.

Venetz and Kasthofer's texts clearly indicate the shift away from Buffon and late-eighteenth-century historical climatology: the real history of the climate, the scene where the major variations unfolded, was now played out in geological time: Kasthofer referred to 'millions of years' and Venetz to the 'mists of time'. This discordance of human and planetary timescales, as well as the glaciers' gigantic size, cast doubt on the possibility of a role for human action in such major upheavals. As for forest cover, Venetz noted that its variation was much too small to yield changes in temperature capable of creating the titanic glaciers of the remote past.

Venetz's paper is often presented as the first formulation of the theory of ice ages. In fact, Venetz, like Kasthofer, was far from envisaging an ice cap covering much of Europe.[46] Even so, it was after reading these works that another Swiss scientist – Louis Agassiz – really did elaborate

44 On the Val de Bagnes affair, see d'Arcy Wood, *L'Année sans été*, pp. 175–95; Ignace Marietan, *La Vie et l'œuvre d'Ignace Venetz, 1788–1859*; and Stefan Berchtold, *Ignaz Venetz im 1788–1859, Ingenieur und Naturforscher*, Brig: Rotten-Verlag, 1990.

45 Ignace Venetz, 'Mémoire sur les variations de la température dans les Alpes de la Suisse. Par M. Venetz, ingénieur en chef du canton du Valais. Rédigé en 1821', *Denkschriften der allgemeinen schweizerischen Gesellschaft für die gesamten Naturwissenschaften*, 1833, pp. 1–38.

46 For example, in the Fiesch glacier, the gap of 2,100 metres between the glacier and the moraine seemed to Venetz almost incredible, given that it would represent an 'inordinate glacier'.

that theory (see Chapter 14). Ice ages were thus a remote consequence of Tambora, and the desire of Swiss scientists to reassure their compatriots that Switzerland was not in the early throes of the thermal death of the globe heralded by Buffon.

A Climate of Laissez-Faire

In France, as in England or Switzerland, the rotten summer of 1816 (in Paris it rained for twenty-six days in July)[47] aroused disquiet about the stability of the natural order. It was then that the idea of a 'disruption of the seasons' took root in French public opinion. Following Toaldo and Lamarck, some explored an astro-meteorological causality;[48] others invoked the presence of icebergs cooling the Gulf Stream, or predicted the death of the Sun.[49] *L'Ami de la religion et du roi*, an ultra-royalist journal, deplored the way all 'these systems about the end of the world' overlooked God:[50] the disastrous weather of 1816 was obviously divine punishment for the Revolution, and should afford an opportunity for national repentance. The rotten summer was also the subject of less serious reflections. On 7 August 1816, under a gloomy sky, Parisians flocked to the Théâtre des Variétés to see a performance of *La Fin du monde ou les Taches dans le soleil*. This vaudeville dramatized discussions about the weather and evoked several apocalyptic scenarios: 'the sun that would go out like a candle', 'spoiled' and eroded by its spots; or an especially malign lunar-solar configuration.

Science was mocked for its certainties in the play. A scientist named Désastres (disasters) gloats smugly: 'Here we have several springtimes spirited away, several summers deferred . . . People everywhere were saying M. Désastres is an imbecile, M. Désastres is an alarmist. . . . Well, this time . . . I stand at my window and watch the world go by; it should make for a picturesque view.'[51]

---

47 'Résumé des observations météorologiques de l'année 1816', *Annales de chimie et de physique*, 1816, vol. 3, p. 443.

48 *Journal des débats*, 21 October 1817; François-Antoine Rauch, *Régénération* de la Nature végétale, ou recherches sur les moyens de récréer dans tous les climats les anciennes températures et l'ordre primitif des saisons par des plantations raisonnées, Paris: Didot Aîné, 1818, vol. 1, p. 264.

49 *Journal des débats*, 18 May 1818.

50 'Du dérangement de la saison, et de la fin du monde', *L'Ami de la religion et du roi*, no. 207, 3 August 1816, p. 369.

51 Lafortelle, Brazier and Merle, *La Fin du monde, ou Les Taches dans le soleil*, Paris: Huet-Masson, 1818, p. 16.

Six months later, during the 'lean season' of spring, the climate question took a more dramatic turn. The consequences of Tambora were now being clearly felt in France: the overall high price of grain and, in the east, a veritable dearth of it, fomented a climate of insurrection. The government's laissez-faire food policy was challenged by many mayors and even some prefects, endeavouring to secure priority access to grain for their citizens. The liberal project of a national economic space unified by market mechanisms collided with the popular moral economy.[52] The government, which opted to defend free prices and trade (by escorting convoys or sending troops to guard markets), took a considerable political risk. Louis XVIII appeared to have abdicated his duties as the nurturing king, preferring to supply the occupation forces that had put him back on the throne rather than cater to the needs of his people. The effect was all the more disastrous in that the population still remembered the rigorous measures taken by Napoleon's government during the shortages of 1811–12. An ultra-royalist critic affirmed that the Restoration government, with its inept laissez-faire, had made itself responsible for 'an extermination of human beings [some 300,000, the author reckoned] no less frightful than the Moscow campaign.'[53]

To justify its liberal policy, the government tried to reconcile two seemingly contradictory figures: the national grain market and the fatherly, nurturing king. In official circulars, respect for the market became a sign of belonging to the national community: 'France is one whole and its inhabitants one family . . . want on the one hand, overabundance on the other, would result from the error which the government would commit by tolerating infringements of free circulation.'[54] Or again: 'The French are the sons of one and the same father, subjects of the same king, and it is only right that objects of consumption should be common to them.'[55] The patriotic discourse of a national market alleviating local penury ultimately rested on the assumption of a fertile, temperate France. France was the liberal land par excellence. Thanks to its fair climate, penury was impossible – *so long as the market was respected.*

---

52 Nicolas Bourguignat, 'Libre commerce du blé et représentations de l'espace français. Les crises frumentaires au début du XIXe siècle', *Annales. Histoire, Sciences sociales*, vol. 56, no. 1, 2001, pp. 125–52.

53 Pons de Villeneuve, *De l'agonie de la France*, Paris: Méquignon, 1835, p. 210.

54 Quoted in Bourguignat, 'Libre commerce du blé', p. 141.

55 Quoted in Pierre-Paul Viard, 'La disette de 1816–1817, particulièrement en Côte-d'Or', *Revue historique*, vol. 159, no. 1, 1928, p. 106.

In the government's view, meteorological anxieties were part and parcel of the rumours that could provoke panics, the hoarding of grain, and eventually an 'artificial shortage'.[56] In official discourse, it was not the natural order that disturbed the social. It was social disorders, disrupting the grain market, that created the illusion of a climate cataclysm. It was therefore crucial to firmly control representations of the weather. The discrepancy between popular meteorological writings and official discourse is striking: whereas commonplace books, agricultural registers or almanacs stressed the disastrous meteorological conditions of summer 1816, Interior Minister Lainé minimized the risks.[57] Neither *Le Moniteur universel*, nor the *Journal de Paris*, enlarged on the rotten summer despite its predictable economic repercussions.[58]

After the satisfactory harvest of summer 1817, and as the shortages triggered by the explosion of Tambora ended, a second theatrical piece, *L'Heureuse Moisson*, drew the political moral from the crisis. The character of Rapinard, a classic ancient-régime monopolist, is strikingly presented as a *meteorologist* – glued to his barometer, eagerly looking out for signs of storm or hail. At the end of the play, Rapinard, who hoarded grain in 1816, is ruined by the following year's good harvest.[59] The moral of the story was that the good citizen must leave it to the laws of the market and natural equilibrium. Seasons of hardship were succeeded by good harvests. On the same model, the government advocated for stoical citizens trusting to both climate and market. Interior Minister Lainé thus presented the good property-owner as someone who 'feared neither difficulties . . . in years of scarcity nor great price depreciation in times of plenty.'[60] Liberal government required a people calm and serene in the face of bad weather.

---

56 'Extrait d'un rapport fait au roi par M. Lainé', *Annales de l'agriculture française*, vol. 1, 1818, p. 366.

57 Cf. Guéneau, 'La disette de 1816–1817', p. 27.

58 A Briard peasant summarized the year 1816 as follows: 'Rainy winter. Late, cold spring. Rainy, late summer. We began to cut the wheat on 20 August. Because of the continual rain, we could never cut for two days running': quoted in ibid., p. 22. In the Nivernais, a joiner noted that 'from May 1816 until 22 September the rain fell without respite; ryes and wheats did not ripen; the grains sprouted inside the sheafs': quoted in Guy Thuillier, 'Les transformations agricoles en Nivernais de 1815 à 1840', *Revue d'histoire* économique *et sociale*, vol. 34, no. 4, 1956, pp. 426–56.

59 Jean-Toussaint Merle, Garmouche and Frédéric de Courcy, *L'Heureuse Moisson ou le Spéculateur en défaut*, Paris: Barba, 1817.

60 'Extrait d'un rapport fait au roi par M. Lainé', p. 367.

In France, as in England or Switzerland, science, too, made its voice heard to reassure the public. The rising star of French astronomy, François Arago, intervened in late 1818 to show that 1816 was not the prodrome of a more profound climate change.[61] His article was presented as a response to debates in Great Britain, but it was primarily addressed to a French audience. Like Leslie, and before him Pilgram and Toaldo, he focused his attention on heavy winters. He drew up a list from 400 to 1740 AD and also used a series of average temperatures in London since 1774. The series of 'heavy winters' demonstrated the existence of very severe cold spells in the past, while London's temperatures evidenced the absence of any multi-decennial trend towards cooling. 'We have no reason', concluded Arago, 'to suppose that the climate of Europe has deteriorated.'

Among contemporaries of the year without a summer, only one, to our knowledge, succeeded in penetrating its secret: the Comte de Volney. Somewhat forgotten in our day, he was celebrated at the time for his career as a traveller, politician, and man of letters. When he took up his pen in 1818, it was, like other scientists, in order to reassure the public. The year without a summer, he explained, was not the sign of some enduring climate imbalance. It was simply an accident attributable to the explosion of Tambora.[62] Volney's perspicacity reflected a broader phenomenon, fundamental for future advances in meteorology: the explosion of the Indonesian volcano coincided with the reconstruction of Europe's mediatic and scientific sphere after Waterloo. Bourgeois, aristocrats, and naturalists took advantage of the return of peace to travel across the continent; periodicals reported on the meteorology of other countries. The year without a summer was thus immediately understood to be a European-wide phenomenon at least. And, if Volney managed to make the link between the summer of 1816 and Tambora, it was because, as an Orientalist, he probably subscribed to the *Asiatic Journal*, one of the few periodicals to have published accounts of the explosion by English officers posted to Java.

It was also in reading the journals of summer 1816 that Heinrich Brandes, a physics professor at Breslau, hatched the revolutionary project

61 François Arago, 'Sur la prétendue détérioration du climat de l'Europe', *Annales de chimie et de physique*, vol. 9, 1818, pp. 292–303.

62 Constantin-François Volney, 'Au rédacteur du *Mercure*, 14 Décembre 1817', *Mercure de France*, vol. 5, January 1818, pp. 111–17.

of *mapping* atmospheric movements. 'From various journals', Brandes explained, 'I had assembled reports on the unique meteorological circumstances of last summer. . . . What is remarkable is that the month of July was rainy, changeable and cold not only in Germany and France, but even in Naples.' He went on: 'If we could illuminate maps of Europe for all 365 days of the year, in accordance with the appearance of the weather, our eyes would see the boundaries of the great rain cloud that covered the whole of Germany and France in the month of July'. For that, he wrote, it would be necessary to set up a continental meteorological network extending 'from the Pyrenees to the chain of the Urals'. And he concluded that 'we would obtain something absolutely new.'[63] Tambora thus laid the groundwork for a fundamental change in scientific approaches to the atmosphere. Brandes's projects of atmospherical cartography would only be implemented from the 1840s, in the wake of research by the astronomers John Herschel and Adolphe Quetelet. But the emergence of a modern meteorological rationality also occurred in the shadow of Tambora, in the search for explanations when confronted with disasters from above.

63 'Extrait d'une lettre du professeur Brandes sur des cartes météorologiques', *Bibliothèque universelle des sciences, belles-lettres et arts*, vol. 4, 1817, p. 264; Heinrich W. Brandes, *Beiträgen zur Witterungskunde*, Leipzig: Barth, 1820. Brandes was not to realize his project for summer 1816. But in 1820 he published a book examining the weather in Europe in 1783 day by day, thanks to data taken from the collections of Louis Cotte and the Meteorological Society of the Palatinate. But he did not deem it right to publish the maps (or was prevented from so doing), which (apparently) guided his reasoning. It was in a thesis in Latin, published in 1826, that he eventually offered the very first spatial representation of an area of pressure at a given date and hour.

9

Should the National Forests Be Sold?

If it was in France, under the Restoration, that the issue of climate change became an affair of state, this was because the effects of Waterloo and of Tambora, the price of defeat and the food shortages of 1816, all intersected there. In the chilly years following the fall of the Empire, Restoration governments proposed to sell the national forests in order to bolster the state's credit and meet the victors' financial demands. The climate issue became entangled with fundamental objectives: to pay war reparations as quickly as possible in order to secure the departure of the occupying troops; to reassure annuity holders and thus rally them to the new regime; to mollify foreign powers and, implicitly, to ensure their support for the Bourbons in the event of new revolutionary disorders.

The option of selling was above all political. Other ways of paying off public debt (compulsory loans, land taxes, indirect royalties, or customs tariffs) existed and were applied. But, by dint of its solemn, spectacular character, the sale of the national forests had the advantage of reassuring capitalists and the buyers of national assets. It was a 'security', a 'mortgage', a 'shield' to protect rentiers from the woes of the public treasury. A political pact was sealed over the fate of the national forests, linking the Bourbons to the new rentier classes. 'The more the mass of the population has an interest in public funds, the more a government is sheltered from revolutions,' declared Talleyrand.[1] The sale of the national forests would

1 Alya Aglan, 'L'invention de la Foi publique (1816–1838)', in Aglan, Michel

restore confidence and make it possible to base the state's credit on the nation's natural, landed and vegetable wealth.

But in both the Chamber of Deputies and in that of the peers, ultra-royalists resisted. From 1814 to 1820, they brandished the climatic argument in every debate on the finance bills: the Revolution had wrecked the French climate and the Restoration must not exacerbate that disaster. At stake in the climate debates over the sale of the national forests lay something fundamental. It was much more than the state budget or the price of defeat: it was the legacy of 1789 and the nature of France.

Forests, Debt and Climate

When, in the spring of 1814, Louis XVIII returned from exile, the kingdom's financial situation was fragile. Baron Louis, the finance minister, assessed the debt at 759 million francs. That is why the new government's first finance law, voted on 23 September 1814, provided for the sale of 300,000 hectares of national forests (out of the 1.2 million owned by the state). These sales did not affect woods belonging to nobles who had emigrated come the Revolution, which were in the process of being returned to their owners. In the main, they involved woods of ecclesiastical origin (dissolved orders, abbeys, chapters), or that belonged to foreign nobles prior to the Revolution. The yield of this property sale was to serve as security for a loan. Alienation was 'the security required for confidence'.[2]

Opposition to the law came chiefly from ultra-royalists, who were surprised and disappointed to see the monarchy honouring Bonaparte's debts. Were the ecclesiastical woods usurped by the Revolution really going to be used to reimburse suppliers of army munitions and 'speculators' who had made a fortune out of the war? Somewhat reluctantly, the ultras defended the national ownership of forests, for it held open the possibility of full restitution in the future. In August 1814, several orators spoke against the plan in the National Assembly.[3] There were numerous

Margairaz and Philippe Verheyde, eds, *1816 ou la genèse de la Foi publique. La fondation de la Caisse des dépôts et consignations*, Geneva: Drox, 2006, p. 80.

2 *Archives parlementaires*, vol. 12, session of 22 July 1814, p. 184; 1 September 1814, p. 550.

3 Michel Bruguière, *La Première Restauration et son budget*, Geneva: Droz, 1969, p. 110f.

objections, centring above all on the danger of selling off highly valuable mature forests. Would not the sudden sale of 300,000 hectares devalue the forestry heritage? The depressed economic circumstances seemed unpropitious for offloading great masses of national woods.

The climate issue reared its head in the debates. Thomas Philibert Riboud, for instance, a deputy from the Ain who was a jurist and agronomist, lauded the ecological virtues of ecclesiastical and seigneurial properties. The clergy, he explained, could not sell its forests (so-called mortmain assets). It therefore of necessity managed them in accordance with a long-term logic. That was why the most beautiful mature woods were to be found on its estates. The aristocrats, for their part, bequeathed them entirely to the (sole) title holder. Everything had been turned upside down with the equality between legatees introduced by the Revolution. For Riboud, this would lead to a division of the woods. Small forestry areas being proportionately more expensive to enclose and monitor, such a division must inexorably lead to deforestation.

In the offing was a frightful climate future that Riboud dangled by recalling the fate of Syria's ancient cities, once brilliant but now lost in the desert. Here was a replay of the account, at once declinist and Orientalist, favoured by Buffon, Bernardin de Saint-Pierre and Rougier de La Bergerie. Dufort, a deputy for the Gironde and himself a former forestry administrator, added to this climatic argument the risk of floods. 'The Pyrenees were once covered in trees', he proclaimed at the tribune; but 'towards the middle of the last century, they were felled on a massive scale. Since then, the Garonne has threatened the beautiful plains bordering it with terrible devastation'.[4] For these two deputies, the forest's ecological function was self-evident and well-known. Given that the same argument had been employed for thirty years, Dufort saw no reason to insist 'on what everyone feels'. Likewise, Riboud contrasted 'those for whom there is little relationship between physical facts and financial arrears' and 'public opinion, which will probably judge otherwise'.[5] Their joint appeal to the natural sensitivity of public opinion served, by contrast, to denounce the Byzantine calculations of financial mechanisms of debt amortization proposed by the government.

Despite opposition from the ultras, the finance law of 1814 allowed for the government to sell up to 300,000 hectares of woods. This law

4 *Archives parlementaires*, vol. 12, 31 August 1814, p. 528.
5 Ibid., 29 August 1814, p. 457.

represented a spectacular break with the principle of the inalienability of the forestry estate – a principle that had been respected during the revolutionary period.[6] Its effects were not long in coming: holders of capital were satisfied, and the annuity rate at 5 per cent leaped from 63 to 78 francs. Meanwhile, for the ultras, the sale of Church woods was an abuse of power. Sacrificing ecclesiastical forests to settle the usurper's bill seemed inconceivable to them. Some ultra-royalist prefects downright refused to proceed without an explicit order from Louis XVIII.[7]

'The Torch of Reason in Our Sacred Woods'

The sales decided in 1814 were to prove insufficient. Following the episode of the Hundred Days (March–June 1815), the kingdom's financial situation worsened considerably. To the arrears of 759 millions, the second Treaty of Paris (20 November 1815) added 700 million francs by way of reparations. Northern France was occupied by an army of 150,000 men, which was a heavy burden on the budget (3 francs per soldier per day). Finally, the allied powers rejected the French default of 1793: the government had to settle an additional 320 million francs, corresponding to the public debt held by private individuals abroad. Ultimately, more than 1,900 million francs were to be transferred to foreign powers during the first decade of the Restoration, representing roughly two years' state expenditure.[8] The matter of a loan, and the consequent restoration of France's credit, became central in political life.

The new finance minister, Corvetto, counted more than ever on the sale of the forests. He envisaged using this sizeable asset (at a conservative estimate the forests were worth 800 million francs) as security for a loan

6 Jacques-Joseph Baudrillart, *Traité général des eaux et fôrets, chasses et pêches*, Part 2, vol. 1, Paris: Huzard, 1823, p. 784.

7 For example, Joseph Fiévée, a famous ultra-royalist journalist who presided at the *Conservateur*. Prefect of the Nièvre from 1813–15, he justified his refusal as follows: 'In the highly unlikely hope of attracting to the King's government the despoiling, revolutionary part of the nation, we are distancing from it all those who remained attached to this government throughout perils and sacrifices.' Cf. Antoine Calmon, *Histoire parlementaire des finances de la Restauration*, Paris: Michel Lévy, 1868, vol. 1, p 102.

8 Eugene White, 'Making the French Pay: The Costs and Consequences of the Napoleonic Reparations', *European Review of Economic History*, vol. 3, 2001, pp. 337–65; André Nicolle, *Comment la France a payé après Waterloo?*, Paris: Boccard, 1929.

raised in London. It is possible that the plan originated in the City itself. According to Nicholas Vansittart, the British chancellor of the Exchequer, as early as 17 September 1815 (even before the signature of the second Treaty of Paris), London banks approached the new government to propose a loan *secured against the national forests*.[9] In any event, the sale was strongly encouraged by the British government. Foreign Secretary Lord Castlereagh wrote in a memo to Wellington that they must be uncompromising with the French state, which could easily afford the reparations: less indebted than Great Britain, it also possessed the treasure of the national forests.[10]

For the French government the situation was untenable, because the elections of August 1815 returned a majority of aristocrats to the Chamber (the famous 'unobtainable chamber', in the words of Louis XVIII), who made the national forests into the banner of their struggle. To restore the clergy to its rightful place, should not its property be returned to it? And to bail out the state, would it not be better to bring back the sale of offices? Faced with the Chamber's intransigence and the fall in the price of annuities, and pressed by Wellington, the king finally decided to dissolve the unobtainable chamber.

In January 1817, the new liberal government, led by Richelieu and Decazes, proposed a finance law to a Chamber it controlled. This law went far beyond earlier planned sales by deciding to transfer *all the national forests* to a new sinking fund: nearly 1.2 million hectares, or 2 per cent of the country's surface area and 20 per cent of its forest cover. According to the budget spokesman, 'the impact of this measure on state credit will be decisive.' The key mechanism in this law was the creation of what would become the Caisse des dépôts et consignations (Deposits and Consignments Fund). More than a mere financial mechanism, the Caisse turned into a 'mediating power',[11] separating what pertained to state credit from what pertained to government policy. A clause in the law afforded protection against the appetites of the Treasury: 'There will not, in any instance

9 Letter from Vansittart to Castlereagh, 19 September 1815, in *Correspondence, Despatches, and Other Papers of Viscount Castlereagh*, London: John Murray, 1853, vol. 3, p. 21.

10 'Memorandum from Lord Castlereagh on the Means of Increasing French Revenue', *Supplementary Despatch and Memoranda of Field Marshal Arthur Wellesley, Duke of Wellington*, vol. 11, London: John Murray, 1864, p. 192.

11 *Archives parlementaires*, vol. 18, 24 January 1817.

or on any pretext, be any infringement of the endowment.'[12] The national forestry capital was henceforth a shield protecting creditors against budgetary jolts; it was the bedrock guaranteeing the nation's credit at all times.

For the ultra camp, this innovation was all the more shocking in that it was intended to attract foreign finance. In January 1817 the bankers Baring of London and Hope of Amsterdam subscribed to a French loan of the order of 300 million francs. The link between the loan and the sale of the woods was obvious,[13] and this scandal united the ultra group in the Chamber. 'Since the woods were assigned to the sinking fund,' wrote Villèle, one of its leaders, 'not one member of this minority has voted for any of the details of the law.'[14]

Between January and March 1817, all the leading lights of the ultra-royalist party took turns to speak against the measure, which they compared to the worst excesses of the revolutionary period: 'The sudden, instantaneous, irrevocable alienation . . . of the forests in their entirety is one of those gigantic projects similar to what during the revolution they called great measures, that were supposed to save France but merely consummated its ruin.'[15] Parodying revolutionary rhetoric, a deputy mocked the 'genius of the century who, torch of reason in hand, enters our sacred woods . . . he no more counts trees for the prosperity of the Treasury than he counted heads for the safety of the nation: he takes the risk of error upon himself.'[16]

The Revolution's Environmental Legacy

The ultras' discourse was informed by numerous arguments. For example, they condemned the treatment of the poor, whom the sale of the national forests would deprive of user rights in order to secure the capital of rentiers. Most importantly, however, as in 1814, their line of

12 Finance Law of 28 April 1816, article 115.

13 Villèle explained it in his correspondence: 'The English government itself supports the loan: it is nearly complete, on conditions that are highly onerous for us. It demands that all the woods be given to the sinking fund. . . . This is what has caused public stocks to rise in the last two or three days.' Letter from M. de Villèle to Mme de Villèle, 10 January 1817, *Mémoires et correspondances du comte de Villèle*, Paris: Perrin, 1888, p. 164.

14 Ibid., p. 209.

15 *Archives parlementaires*, vol. 18, 7 February 1817, p. 628.

16 Ibid., vol. 18, 14 January 1817, p. 157.

argument foregrounded the disastrous *ecological* effects of current policies – and the consequences of Tambora, of course, made climate a very auspicious theme. For the ultras, the revolutionary events, as well as disrupting the political, religious, social, and moral equilibrium of France, had jeopardized its natural equilibrium. These terrible blows to French nature, hitherto preserved, maintained, and cherished by generations of monarchs, aristocrats and servants of the Crown, had especially affected the kingdom's forests and consequently its climate.

Thus, in France, according to Adrien de Rougé, 'arid rocks, deep ravines, everywhere replace forests . . . in several places springs have dried up, and droughts *unknown before the revolution* now often scorch the products of the soil' (our emphasis).[17] Castelbajac, for his part, took the example of departments of the Midi, 'annually devastated [by storms] ever since stripped mountain summits attested to *the passage of a revolution*'.[18] The deputy Aurran-Pierrefeu homed in on his own department of the Var, where 'tree deaths have become more frequent and more widespread *since the revolution*, as a result of the clearing of mountains and the destruction of the greater part of the forests.'[19]

By honouring illegitimate debts and, what's more, doing so by alienating forests stolen from the Church, the government was perpetuating the destruction visited by the sans-culottes of Year II, the purchasers of national assets and the suppliers of the Grande Armée. In linking the credit of an impecunious state to the forests, the sinking fund transformed financial into ecological profligacy. 'If France had an enemy intent on her ruin,' wrote Louis de Bonald in the ultra journal, *L'Ami de la religion et du roi*, 'since he could not drain the seas, or strip her soil of its natural fertility, or her air of its salubriousness, he would have her forests sold.'[20] The ultra deputy Josse-Beauvoir linked alienation to the post-Tambora food shortages, which allowed him to underscore the injustice of a law designed at once to enrich capitalists (by raising the annuity rate), damage the climate, raise cereal prices, and cause a 'loving, loyal people' to suffer.[21]

We find the same arguments in the Chamber of peers. The Marquis de Louvois duly recapitulated the 'physical dangers' of alienation: the threats

17 Ibid., vol. 16, 14 March 1816, p. 546.
18 Ibid., vol. 18, 5 February 1817, p. 545.
19 Ibid., vol. 19, 22 February 1817, p. 83.
20 *L'Ami de la religion et du roi*, 4 March 1817, vol. 11, p. 120.
21 *Archives parlementaires*, vol. 19, 4 March 1817, p. 285.

to France included erosion, hurricanes, aridification of the soil, and the disappearance of springs and vineyards. No doubt addressing the astronomer Pierre-Simon de Laplace, the strongman of Parisian science, who sat with the peers, Louvois called for confirmation of these effects: 'ask [those profound scientists] who sit amongst us and they will tell you, like me, that a forestry mass is indispensable to France's health and fertility.'[22]

Chateaubriand, too, the most brilliant pen of the ultra party, rose to the defence of the national forests and the climate of France. From the heights of his experience as a traveller, the author of *Natchez* and *Itinéraire de Paris à Jérusalem* affirmed the reality of the process that led from deforestation to the end of the world. 'Wherever trees have disappeared, man has been punished for his improvidence: I can tell you better than most, Messieurs, what the presence or absence of forests leads to, since I have seen the loneliness of the New World, where nature seems to be being born, and the deserts of old Arabia, where creation seems to expire.'[23]

Chateaubriand further rejected alienation as a short-termist decision: 'I know that in this century people are little moved by reasons lying beyond the term of their existence: everyday misfortune has taught us to live from day to day. . . . Time flies rapidly in this country; in France the future is always near at hand.'[24] This was another facet of the discourse of the ultras, who denounced the sale of the national forests as an attack on 'future generations'. 'What generation is it', Bonald demanded, 'that can arrogate to itself the right to dispose of land belonging to every generation of Frenchmen?'[25]

The government party responded to the ultras' ecological arguments in two ways. Firstly, by uncoupling the privatization of national forests from their destruction. The market would perform the role of regulator: if buyers dared too much deforestation, the depreciation of wood, suddenly in excess, would soon restore them to their senses. Secondly, by insisting that the link between deforestation and the decline of nature was anything but proven. Étienne Pasquier, the justice minister, reminded deputies of the teaching of Genesis: 'It was everywhere given to man to change the face of the earth he inhabits.' This was a credo which climate

22 Ibid., 22 March 1817, p. 559.
23 Ibid., 21 March 1817, p. 498.
24 Ibid., 21 March 1817, p. 503.
25 Ibid., 4 March 1817, p. 264.

warming since antiquity seemed to confirm: 'Was the harsh climate of the Gauls in Caesar's time . . . preferable to the climate of France today [?]'[26] The budget spokesman, Jacques Beugnot, emphasized the uncertainty of forestry and climate data: 'Has the destruction of woods really been so great in France over the last forty years?' 'Remember, messieurs, that three years ago the then constant dryness of our summers was wont to be explained by the decline in forests. Scientists and distinguished writers said clearly that we should no longer hope for more than a few drops of water in the summer; last year's summer proved them sadly wrong.'[27] The ultras' climate fears were belittled as a matter of romantic 'sensitivity'. Camille Jordan expressed surprise at the 'frivolous complaints' of those whose 'hearts seem to have contracted a sort of chivalrous enthusiasm for these noble trees.'[28]

In the early years of the Restoration, two visions of France were basically arrayed against one another through the prism of forestry and climate. The first, articulated by successive governments, defended a France that was a financial and commercial power. The example to be followed was obviously England, which, although stripped of its woods, maintained an enormous navy for war and trade. As early as 1814, Finance Minister Baron Louis stressed the capacity afforded by credit and financial power to *externalize* the forestry issue, enabling resources to be purchased abroad: 'the sea, together with credit, would accumulate more naval timber in our ports than could be produced by all our forests.'[29] In the Chamber of peers, Count Couteulx de Canteleu, one of the kingdom's foremost bankers and regent of the Bank of France at its inception, favoured the sale of national assets, for 'it is no longer on our cannon that we should inscribe the adage adopted during the reign of Louis XIV, *ultima ratio regnum* . . . today this inscription should be placed on the Royal Treasury.'[30]

A contrasting view of France was articulated by the ultras: a France of the land, resting above all on its national resources and thus on the quality of its nature and climate. This was a more embodied France, conceived in the variety of its regions, with its strengths and weaknesses. It meant Riboud, for example, challenging the case for the replacement of wood by

26 Ibid., 22 March 1817, p. 520.
27 Ibid., 5 March 1817, p. 301.
28 Ibid., 4 March 1817, p. 271.
29 Ibid., vol. 12, 2 September 1814, p. 581.
30 Ibid., 20 September 1814, p. 699.

coal, since deposits of the latter only existed in some parts of the country. This view refused to consider international trade as a normal phenomenon and regarded the forest as so much emergency capital, mobilizable in the event of war and autarky. For Bonald, public credit was not as important as it seemed: unlike England, France was a vast country with large agricultural areas that were still amenable to improvement. It was with this essential objective in mind that capital should be mobilized, not to pay off debt. Spain's heroic resistance to Napoleon's armies proved the point: a nation could defend itself without credit and even with practically no money.[31]

Since 1790, the forestry estate had been governed by a principle of inalienability that had been effective in shielding it from massive sell-offs. This changed utterly with the fall of the Empire. In the early 1820s, Restoration governments would offload some 169,000 hectares of national woodland – an enormous loss to the forestry estate.[32] How are we to explain this volte-face? Despite their passionate resistance, the ultras did not stand a chance. Until 1830, France numbered fewer than 100,000 electors and nearly double that number of rentiers.[33] This means that *all* active citizens under the Restoration had a direct interest in the sale of national forests which, by raising the value of annuities, swelled their assets. Alienation was the well-nigh logical consequence of the censitary regime and the power of rentiers. The public debate and the ultra-royalist parliamentary guerrilla nevertheless left an important legacy. The 1817 finance law imposed a vote in Parliament as the precondition for every alienation. Every year, the forestry debate was relaunched and every year climate change was on everyone's lips.

31 Antoine Calmon, *Histoire parlementaire des finances de la Restauration*, p. 164.

32 Jules Clavé, 'L'aliénation des forêts de l'État', *Revue des Deux Mondes*, vol. 62, 1866, pp. 197–214. The loss was the greater in that the forest surface area, difficult to measure precisely, sold back to nobles returning from exile in 1814 must be added to these sales.

33 A. Legoyt, *La France et l'Étranger: études de statistique comparée*, vol. 1, p. 246.

10

The Crusades of François-Antoine Rauch

In the decades following the Revolution, political assemblies, scientists' offices and ministerial corridors were not alone in resonating with warnings about the plight of the French climate. Newspapers and magazines also broadcast these alerts, propagated visions of a pristine or ruined nature, and mobilized political camps that accused one another of having destroyed the seasons.

A whole journal was devoted to the issue of anthropogenic climate change in the 1820s. It was created by the determination of one man, François-Antoine Rauch, who succeeded for a while in capturing something of the spirit of the times and transmuting it into an often emphatic prose, sometimes muddled but invariably inspired. His *Annales européennes* appeared from 1821 to 1827. They were the crucible of some of the most influential tropes in Western discourse about nature and its deterioration.

The years of the Restoration, when the effects of Tambora and bitter struggles over the environmental legacy of 1789 merged, elevated him to the rank of a public figure.[1] But Rauch intended to be more than just

1 Raphaël Larrère, préface to *L'Harmonie hydrovégétale et météorologique, ou l'Utopie forestière* (extracts from Rauch's text), Rungis: INRA, 1985; 'L'harmonie hydrovégétale et météorologique: l'utopie forestière de Rauch', *Milieu*, no. 21, 1985, pp. 40–45; Jean-Baptiste Fressoz and Fabien Locher, 'Régénérer la nature, restaurer les climats: François-Antoine Rauch et les *Annales européennes de physique végétale et d'économie politique*, 1815–1830', *Le Temps des médias*, no. 25, 2015, pp. 52–69; Ford, *Natural Interests*, pp. 16–42; Guillaume Decoq, Bernard Kalaora and Chloé Vlassopoulos, *La Forêt salvatrice: reboisement, société et catastrophe au prisme de l'histoire*, Seyssel: Champ Vallon, 2016, chapter 3.

a whistle-blower. In a second life, less well-known than the first, he also sought to interest the world of business in his climate crusade. He called on capitalists to transform the face of France and accomplish a major restoration of its natural equilibria under his guidance. The climate prophet was also, and perhaps above all, an entrepreneur: of himself, of his climatic cause, but also of financial packages profiting both.

Our era has celebrated him as a precocious environmentalist for his odes to a nurturing nature, to respect for balance, to harmony. In him, we find side by side a love of technology, a thirst for celebrity and an attraction to money. Above all, he had an ardent faith in man's ability to repair nature.

Rauch's Vision: A Material, Global and Divine Harmony

Amid the turbulence of the 1790s, anything seemed possible. The task was immense: regenerating a country, a people, a nature, after centuries of tyranny. Rauch was thirty and wanted to be part of it. For him this meant conceiving and administering grand projects. Here he was not particularly original. Following the Revolution, scientific and technological utopias proliferated. The new social and political order, it was everywhere proclaimed, must break with centuries of particularism, routine, and backwardness. From units of measurement to public facilities, everything must contribute to the reign of rationality, of universality, and hence of liberty and abundance. Rauch embraced this movement, which chimed with his own professional culture: that of a budding engineer, who shared with his fellow students at the École des Ponts a strong ethos of construction and planning.[2] He was still a student there when, in 1792, he published his first grandiose plan for infrastructures destined to change the face of France. Through a building programme for mills and granaries, this *Plan nourricier* would 'ensure bread for the French people for ever'.[3] Two years later, having left the school and now an engineer in the Meurthe, he sought to promote the digging of canals in order to equip the country with an immense network of inland navigation – something

2 He entered the École des Ponts in 1783, aged twenty-one, as a free auditor. Following posts in États de Liège and Roussillon, he rejoined the École in autumn 1792, this time as titular student.

3 François-Antoine Rauch, *Plan nourricier, ou Recherches sur les moyens à mettre en usage pour assurer à jamais le pain au peuple français*, Paris: Didot Jeune, 1792.

his superiors at the Ponts et Chaussées administration viewed with scepticism.[4] Then, in Year VI, he urged its management to seriously embark on building a canal in Suez.[5] This kind of direct appeal remained his trademark, with public or private addresses to all levels of government. In 1792, for example, he sent letters to the National Assembly, to the king, to ministers. Though concerning his massive projects, these initiatives often had personal motives too. Rauch, whose career as an engineer was chaotic to say the least, and whose agricultural enterprises often failed,[6] frequently advocated for his own cause.

Above all, Rauch was a man with a vision. In March 1792, he wrote to the president of the Assembly to warn against the selling of national forests. He detailed the impact it could have on wood resources, but, even more importantly, on the 'harmony of elements' connecting forests, celestial phenomena, waterways and vegetation.[7] He vowed to return to the subject in a book, which he only finally published in 1802, under the title *Harmonie hydrovégétale et météorologique*.[8]

Rauch's work was the start of a crusade lasting more than thirty years for a forestry and climatic regeneration of France. 'The harmonic laws of Nature, the mother who nurtures all beings, have been distorted by long centuries of mutilations . . . the celestial phenomena unleashed now only disclose themselves in devastating effects.'[9] One man could step in to avert this apocalypse: Napoleon. The frontispiece of *Harmonie hydrovégétale*

4 *Rauch, ingénieur des Ponts et chaussées, ayant l'arrondissement des districts de Dieuze, Sarrebourg et Salin-libre . . . aux agriculteurs, manufacturiers, commerçans et hommes éclairés de la contrée*, Dieuze: n.p., Year III, pp. 10–11.

5 Letter from Rauch to the Executive Directory, 30 Germinal Year VI. Dossier 'Rauch' (AN F14/2308/2).

6 In 1799 the Ponts et Chaussées administration appointed him to a new post, but he refused to leave Meurthe and was placed on inactive status. In 1802 he agreed to becoming an engineer in the Bas-Rhin, but definitively quit the Ponts et Chaussées administration in 1810, aged less than 50, after having rejected a new appointment. At the same time, he pursued agricultural activities by trying to grow tobacco and sugar beet at his property in Vergaville (Meurthe). Report from the head engineer of Meurthe dated 9 Frimaire Year VII. Dossier 'Rauch' (AN F14/2308/2).

7 Rauch, *Plan nourricier*, pp. 112–16.

8 Rauch, *Harmonie hydrovégétale et météorologique, ou Recherches sur les moyens de recréer avec nos forêts la force des températures et la régularité des saisons*, 2 vols, Paris: Levrault, Year X. He presented an outline of his ideas in 1800 to the appropriate minister, Chaptal. Letter from Rauch to Chaptal (Interior Minister), 24 frimaire year IX, Dossier 'Rauch' (AN F14/2308/2).

9 Rauch, *Régénération de la Nature végétale*, vol. 1, p. vii.

represents him as a veritable demiurge, admiring a France repaired by his actions, thanks to Rauch's wise counsel.

The ruptured harmony affected the great water cycles. According to Rauch, every day some 47,000 billion tons of water evaporate from seas, lakes and the ground, to circulate between Earth and the sky. These gigantic volumes return to their original sources in three ways: via rain and snow, through freezing at the poles and on mountaintops, *or by being sucked into the ground by trees*: 'With the attraction of their canopies, from afar they command the wandering waters of the atmosphere to come and pour into their protective urns',[10] he wrote. It is trees that ensure equilibrium and, by returning water to the ground, enable vegetation to flourish and animals to live.

Deforestation disrupts this providential cycle: it increases humidity in the atmosphere, 'and one dreads to think what these suspended seas must become when diminished plants can no longer pump out the half of them.'[11] Once trees are felled, atmospheric water no longer condenses into dew or fertilizing showers, but falls to Earth as downpours, or accumulates in the form of snow and ice at the poles or on mountain tops. Hence a new climatic regime made of droughts and torrential rain, arid ground and catastrophic floods. Hence also, because forests no longer perform their sheltering role, the proliferation of storms and hurricanes.[12] Rauch thought of climate change as a *planetary* phenomenon: by deforesting, humanity had disturbed the water cycle that linked the tropical zones to the poles. The growth of the ice cap since the seventeenth century, diagnosed by Buffon and numerous naturalists in his wake, was (according to Rauch) caused by the deforestation of Europe and America.

His theory drew on two sources. Firstly, he referred to works by the English naturalist Stephen Hales on the physiology of plants, insisting on the importance of the volumes of water exchanged with the atmosphere by plants.[13] Next, he paraphrased Bernardin de Saint-Pierre in stressing that the very morphology of trees was an expression of their role in the providential order. Bernardin's influence was ubiquitous: Rauch claimed to have known him personally, and that Bernardin had supported him at the beginning of his studies on forests and climate.[14]

10 Ibid., p. 8.

11 Ibid., p. 10.

12 Ibid., pp. 60–9.

13 Rauch, *Harmonie hydrovégétale et météorologique . . .*, vol. 1, p. 13.

14 Rauch, *Régénération de la Nature végétale*, vol. 1, p. 30; and letter of 22 May 1817 from Rauch to the Interior Minister, dossier 'Rauch' (AN F14/2308/2).

'Regenerated France is asking you to recreate this beautiful nature over its entire surface.' François-Antoine Rauch, *Harmonie hydrovégétale et météorologique*, 1802 (frontispiece).

But Rauch otherwise mistrusted theoreticians and what he called the 'barren heaths of hypotheses'. He aspired not so much to do science as to make a sensation. His style prized evocative power above precision: trees were 'siphons mediating between clouds and earth';[15] woods became 'aromatic ventilators'[16] that 'cushioned the winds';[17] marshes were 'the putrid sores of the earth',[18] and so on. Although an engineer at the Ponts et Chaussées administration, Rauch refused to adopt the posture of the expert and insisted rather on the importance of a civic science of nature: 'I observe nature not with the science of a naturalist, but with the heart of a good citizen.'[19] He called for the revival of simple forms of knowledge and for contemplation of the 'infinite marvels' of the natural world, a source of 'moral happiness', the better to appreciate and defend them.[20]

15 Rauch, *Régénération de la Nature végétale . . .*, vol. 1, p. 8.
16 *Annales europées*, vol. 3, p. 251.
17 Rauch, *Régénération de la Nature végétale . . .*, vol. 1, p. 24.
18 Ibid., vol. 2, p. 187.
19 Rauch, *Harmonie hydrovégétale et météorologique*, vol. 1, p. ii.
20 *Annales européennes*, vol. 3, p. 10.

Babylon, or The Ruins of the Future

The media had long been buzzing with the echoes of storms, disastrous seasons, floods, abnormal snows or heat. These had filled column inches ever since the advent of the daily press in the late eighteenth century. Under the Restoration, the growing number of local news sheets gave prominence to meteorological news, linking it to harvest forecasts. At the start of the 1820s, one sub-prefect described the anxiety-inducing hype to his minister as follows: 'If, today, storms seem to be more violent and frequent than in the past, the reason is that journalists, who do not always have political or literary material to hand to adorn their daily sheets, employ these sorts of events as padding. In this they appear wonderfully well-served by the scribbling fashion that has become epidemic in our day.' Worse, with journalists happily publishing letters from their subscribers, 'every newspaper reader believes himself obliged to publicize, by amplifying it, the scourge that has just afflicted his own village.'[21] The *Annales européennes* launched by Rauch in 1821 were fuelled by this meteorological public sphere.[22]

The launch took place in an ambience already saturated with worries around climate deterioration: everyone remembered the food riots of 1816–17. With this journal, Rauch set out to convert the patchy material disseminated in the press, as well as in scientific and administrative literature, into an awareness of the human responsibility for climate disruption. The subject was still controversial, after all. For many, humanity lacked the power to adversely affect the major equilibria of the planet.[23] Moreover, before the 1830s, the notion of a meteorology controlled by the stars still survived, even if it was losing ground. As late as 1810, the naturalist Jean-Baptiste Lamarck published an *Annuaire météorologique* in which he popularized a theory of lunar meteorological action.[24] Such

21 Report by J.B. Serre, sub-prefect of Embrun, dated 31 December 1822. Dossier 'Basses-Alpes', Académie des sciences, collection of the inquiry of 25 April 1821. See Chapter 11.

22 See, for example, Rauch, *Harmonie hydrovégétale et météorologique*, vol. 1, pp. 76–80, 104–10; *Régénération de la Nature végétale*, vol. 1, p. 281; *Annales européenes*, vol. 1, p. 363; ibid., vol. 5, pp. 414, 453; ibid., vol. 8, p. 335; ibid., vol. 9, p. 207.

23 See, for example, François Dominique de Reynaud de Montlosier, Chapter 11 of this book.

24 François Petit-Perrin, 'La météorologie de Lamarck', master's thesis, University of Paris I, Panthéon-Sorbonne, 2005.

explanations, which exonerated human action, occasionally re-emerged in the Restoration press.[25] This was precisely what Rauch was fighting against. 'It is not in the region of the stars', he wrote in his *Annales*, 'that we should search for the causes of the ominous changes in climates . . . these causes are close by us and come from us. *It is the hand of Man that weighs upon the Globe*' (our emphasis).[26]

To win converts, Rauch relied on rhetoric, whose power he often lauded. He promised his readers a 'spectacle', 'attractive images', a 'gentle diversion', and even the 'enjoyment' of viewing the 'might of nature'.[27] How were the natural harmonies between forest and climate to be made tangible? How to convey that the matter of forests 'embraces with it all the inhabitants of the Earth, the air and the seas', and that 'everything in nature touches everything else'?[28]

This point in particular was where Rauch innovated, by producing one of the first declinist frescoes of the human history of the natural world. Buffon's Époques de la nature served as an inverted model. In Rauch, as in Buffon, human history and the history of the globe are closely linked and written in tandem. But the key difference attached to Rauch's providentialist and creationist convictions: nature could not be unstable, or rushing headlong to destruction. That is why Buffon's theory of the Earth's thermal death seemed impious to him. Man's action alone was a vector of destruction: all man's actions were open to question, since the world was originally perfect.

Rauch's grand fresco unfolded in three acts and an epilogue.

Act One: primitive nature. From Rauch's standpoint, having a precise view of nature prior to, or at the very inception of, human intervention was crucial for defining principles of good management. Rather than reasoning in the abstract, Rauch explained that it was wiser 'to confine oneself to envisaging the condition in which Creation [had] appeared'.[29] Hence his accent on the pantheism of ancient religions ('which had animated all of nature with their bucolic deities'[30]), and, above all, the presence in

25 Cf., for example, *Journal de débats*, 21 October 1817; Rauch, *Régénération de la Nature végétale*, vol. 1, p. 264.

26 *Annales européennes*, vol. 1, p. 340.

27 *Dictionnaire des sciences naturelles*, Strasbourg and Paris: Levrault, 1824, vol. 30. Cf. the prospectus for the *Annales européennes* published at the end of the book, p. 16.

28 Rauch, *Harmonie hydrovégétale et météorologique*, vol. 1, pp. 62–3.

29 Rauch, *Régénération de la Nature végétale . . .*, vol. 1, p. 108.

30 Rauch, *Harmonie hydrovégétale et météorologique . . .*, vol. 1, p. 56.

his writings of two sites emblematic of the natural Eden: Tahiti and the Amazonian forest.

For Rauch, Tahiti acted as a social and ecological model. The island offered the 'touching image of the ancient beauty of nature'; its inhabitants were 'content with the treasures nature has lavished on them, never dreaming of destroying or diminishing them.'[31] He thus adopted the familiar topos of the Pacific islands as 'paradise islands', but with the peculiarity of underlining not only the benevolence of their nature and climate, but also their integrity as *preserved* by the virtue of their inhabitants. The Amazonian jungle was the other pole of Rauch's Eden: far from regarding it as a humid, unhealthy nature, left unattended by the savages (as in Buffon), he conceived it as a vestige of Creation, as the image of a lost paradise. In the *Annales européennes* he reproduced the then famous engraving by the Comte de Clarac,[32] to stimulate in his readers an early form of ecological nostalgia for primitive forests.[33]

Act Two: North America. Rauch inflected the then dominant ecological understanding of the American continent. Whereas North America was generally seen as the image of Europe's past, affording colonists the opportunities and riches of a nature that was still intact, Rauch portrayed a continent devasted by the 'rush to gratification' on the part of colonists 'from all countries'.[34] In scarcely a century and a half, the lure of profit had already led to the deforestation of a surface area greater than Europe.[35]

Act Three: the Orient. Here, Rauch reproduced Buffon's negative view, adopted by many authors of the time. By dint of his poor management of nature, improvidence and warmongering violence,[36] the Oriental was responsible for the ecological collapse of his territory: woods demolished, springs exhausted, rivers dried up or turned to marsh, climates degraded and insalubrity all of which, in return, caused racial degeneration.[37] 'Nineveh and Babylon . . . the sites of the world's first political storms, no

31 *Annales européennes*, vol. 3, p. 308.

32 Charles de Clarac, a famous archaeologist, had visited the Amazon in 1816. See Pedro Corrêa do Lago and Louis Franck, *Le Comte de Clarac et la forêt vierge du Brésil*, Paris: Louvre-Chandeigne, 2003.

33 *Annales européennes*, vol. 5, p. 140.

34 Rauch, *Régénération de la Nature végétale*, vol. 1, p. 94.

35 *Annales européennes*, vol. 1, p. 116.

36 Ibid., p. 106.

37 Ibid., p. 110.

Biodiversity Heritage Library

'The Virgin Forest of Brazil' (Charles de Clarac), *Annales européennes*, vol. 4, issue 16

Biodiversity Heritage Library

'The last remains of Babylon', *Annales européennes*, vol. 4, issue 14.

longer have any witnesses of their past existence and magnificent ruins other than silent deserts.'[38]

The final scene was played out in Europe. The Oriental collapse must serve as a warning to governments: 'Highly populous Europe is headed with frightening rapidity towards the annihilation of woods and waters whose disappearance has forever withered . . . the most brilliant regions

38 Rauch, *Harmonie hydrovégétale et météorologique*, vol. 1, p. 45 and *Annales européennes*, vol. 1, p. 105.

of the East.'[39] The continent was at a crossroads: if the European powers, finally at peace since 1815, heeded Rauch's solemn message, the disaster could still be averted. Rauch once again resorted to imagery. A series of three engravings displayed the present state of France (denuded mountains, desiccated plains empty of wild animals, absence of fish and desperate fishermen); the enterprise of 'fructification' to be undertaken (nurseries, tree planting, stocking of rivers and streams); and the pleasant results that could be expected (re-wooded slopes, fertile countryside, rivers with fish and happy fishermen: see the images on pp. 151 and 152).

While it is difficult to measure the diffusion of these ideas precisely, Rauch benefited from powerful relays in the press, provincial scientific societies, and even the royalist government. Between 1821 and 1823, most of the major papers reviewed the *Annales* in highly approving fashion: 'An eminently national and, at the same time, European work' (*Le Moniteur universel*);[40] 'This work develops with rousing eloquence everything our Globe offers that is most marvellous' (*La Gazette de France*);[41] 'Embraces the plant and meteorological regime in their highest aspects' (*Le Drapeau blanc*); 'Rises to the highest thinking of Creation' (*Le Courrier français*); 'No scientific and literary enterprise is more worthy of focusing our attention' (*Le Miroir*); 'Profound imprint of the religious spirit that links all physical causes to a supreme power' (*Journal de Paris*).[42] And, if Rauch was practically the only contributor of original texts to his review, it published an ample correspondence with provincial scientific societies.[43] For instance, Rauch had contacts in Provence where deforestation was a very live issue,[44] in Bordeaux, where he was supported by the Linnean Society,[45] and in Saint-Étienne.[46]

39 Rauch, *Harmonie hydrovégétale et météorologique*, vol. 1, p. 47.

40 *Le Moniteur universel*, 8 June 1822.

41 See the prospectus for the *Annales* in Rauch's engineer file (AN F14/2308/2)

42 *Journal de Paris*, 10 March 1822.

43 For example, a scholar from the Vosges, who was a member of the departmental antiquities committee, brought him evidence of the existence of vines in the heart of the Vosges around the year 1000. The climate was said to have suddenly deteriorated in the mid-seventeenth century as a result of a deforestation impelled by the army. *Annales européennes*, vol. 4, pp. 1–17.

44 His theories were taken up by local societies and relayed by *L'Ami du bien*, a philanthropic journal based in Marseilles. *Annales européennes*, vol. 10, p. 64 and vol. 11, p. 136.

45 Ibid., vol. 5, p. 257.

46 *Bulletin d'industrie agricole et manufacturière*, vol. 3, 1825, p. 120.

'The deplorable state of many of France's cantons', *Annales européennes*, vol. 5, issue 19.

Given Rauch's wait-and-see attitude under the Restoration, his exact political position is difficult to define. While, on the one hand, he was clearly aligned with the ultra-royalist stance against the alienation of national forests,[47] on the other, he took great care to present his cause as a 'national' one – that is, 'foreign to political debates'.[48] The Society of Christian Morality took an interest in him because it regarded his exhortations to contemplate nature as a useful way of promoting the faith. The royalist press – the *Journal des débats* – or ultra press – *L'Ami de la religion* – engaged in the struggle against the sale of national woods, also disseminated his ideas.[49] In any event, it was between 1817 and 1821, when the matter of the alienation of the forests occupied the political centre-stage, that Rauch obtained his greatest successes. His writings even acquired a semi-official function: in 1818, the interior minister asked all prefects to promote Rauch's latest book. Mayors were firmly invited to purchase copies of it.[50]

47 He published Louis de Bonald's long speech inveighing against the alienation of the national forests in 1817: *Annales européennes*, vol. 4, p. 23.

48 Ibid., vol. 10, p. vii.

49 In 1819 the geographer Malte-Brun published a long review of *Régénération* in the columns of the *Journal des débats*, congratulating the author on his success with the interior minister, but noting that the Finance Ministry was continuing to sell off the forests: *Journal des débats*, 16 July 1819, pp. 3–4. The same year, the ultra organ *L'ami de la religion et du roi* criticized Rauch while seizing on the political advantages that could be derived from his climate crusade: *L'Ami de la religion et du roi*, vol. 20, 1819, p. 160.

50 Letter of 9 March 1818 from the Comte de Chabrol to the prefect of the Nord. Archives Départementales du Nord, M500/15; *Annales européennes*, vol. 5, p. 52.

'The fructification of France's wastelands and empty waters', *Annales européennes*, vol. 7, issue 25.

'The regeneration of the blessings of nature in all France's cantons', *Annales européennes*, vol. 5, issue 20

The Bad Business of the Climate

Rauch's climate crusade did not consist only in convincing public opinion or alerting the government. He also wanted to act personally and directly on the nature and climate of France by mobilizing other allies: capitalists.

In 1824, he tried to found a Society for the General Fructification of the Land and Waters of France,[51] aiming at nothing less than 'the physical regeneration of the Realm'. It was a grandiose plan. For 100 million francs, Rauch planned to obtain the concession of *all* the uncultivated land belonging to the state and communes (marshes, wasteland, moors) for a period of ninety-nine years, in order to 'improve their value' by draining or clearing them, prior to transforming them into woods or agricultural holdings. Rauch applied for permission to the public authorities, without success: the rules of public limited companies required an initial input of one-quarter of the social fund[52] – and Rauch did not possess a single franc of the 25 million required.

In 1826, Rauch created the Company for the Fructification of the Wasteland and Uncultivated Land and Waters of France on a different model (limited partnership with shares).[53] In the memorandum of association, he indicated that, thanks to the help of prefects, he had assembled a 'mass of land plots available for fructifying'. In reaction, no doubt, to the criticisms prompted by the initial version of his plan,[54] the company also promised some land purchases. Most importantly, the plan had developed in terms of substance: there would be 'cultivation' of uncultivated lands, but there was no more mention of planting woods. The plan remained ambitious enough: 40,000 shares at 1,000 francs, for the exploitation of a million hectares, no less. In September 1827 that was a very distant prospect, but even so Rauch managed to collect 80,000 francs from twenty-five investors including employees, property-owners, doctors and lawyers.[55]

51 *Société de fructification générale de la terre et des eaux de la France, ayant pour but: la régénération généralement désirée et qui peut s'effectuer dans l'espace de dix ans de toute la nature végétale*, Paris: Trouvé, 1824.

52 Letter from the interior minister to Rauch, 13 July 1825, AN F12/6809.

53 He joined forces with Michel Duverdier de Vauprivas, a former military man charged with protocol functions at Court. Act of 26 May 1826 before the notary Benjamin-Victor Vernois (AN ET/III/1431), and act of 28 August 1826 (AN ET/III/1432).

54 'Société anonyme de fructification générale', *Bulletin des sciences agricoles et économiques*, July 1825.

55 Act of 14–15–16 and 18 September 1827 (AN ET/III/1446).

But with the Company failing to secure enough land, Rauch was compelled to refound it as the Company for Drainage and Clearance (of marshes, dunes, wastelands).[56] Among the assets declared under the new structure were the statistics he claimed to have amassed on the uncultivated land of France and the preference it had acquired, in concession agreements, on 74,000 hectares of marshes and ponds. But the operation still did not take off: in January 1828, the minimum quota of 2,000 shares not having been subscribed, no more were issued, signalling the failure of the attempted relaunch.[57]

Rauch's enterprises might seem laughably grandiose. He had, after all, explained that his project was to 'transform France in fifteen years into a terrestrial paradise'.[58] Was he merely one of those semi-megalomanic, semi-charlatan 'project developers' of which the era was full? Not at all: his contemporaries took him seriously, for his projects were perfectly in line with several Restoration movements which he synthesized in an original way.

In the first place, the late 1810s and the 1820s saw a peak in the price of agricultural products, so that 'uncultivated land' aroused the covetousness of many capitalists. When Rauch launched his companies, for instance, a Company for the Clearance of the Moorlands of Bordeaux was already angling for the concession of 50,000 hectares of such land for ninety-nine years in order to build 500 farms on it, with a total investment of 25 million.[59] Rauch's project also coincided with a flurry of capitalist excitement around drainage operations – a trend noticeable at the same time in Italy, England and Holland.

Secondly, while we know that his career was bumpy, Rauch could present himself as an expert: as a former engineer of the Ponts et Chaussées administration, he possessed technical credibility in the world of public works, politics, and business.

Thirdly, Rauch's projects proceeded from a mercantilist current, blending agronomy and acclimatization, which emerged revitalized from the

56 Act of 21 September 1827 (AN ET/III/1441). This time he managed to team up with two influential partners: first Aimé Jourdan, who had been chief clerk in Finances (an important post) at the start of the Restoration; and then Baron Plein, an engineer in Ponts et chaussées and former general officer of engineers. These were associates whose personal status was, on the face of it, sufficient to reassure investors.

57 Act of 15–16 January 1828 (AN ET/III/1441).

58 *Annales européennes*, vol. 6, p. 261.

59 Ibid., vol. 5, p. 33.

continental blockade. Moreover, Rauch had already tried the production of beet sugar, though without success. On numerous occasions he insisted on the patriotic dimension of his companies, which would enable France to forego certain imports.[60] Rauch added a touch of originality: he claimed that his enterprises would be able to recreate the favourable climates of the lost colonies at home,[61] making it possible to grow new crops.

Fourthly, Rauch's project corresponded to a form of agricultural philanthropy rooted in the late eighteenth and early nineteenth centuries (as in Boncerf, for example; see Chapter 6). Paupers and beggars would come to settle on the regenerated lands. The *Annales* published a long article on the charitable Society founded in 1818 by Prince Frederick of the Netherlands, reporting that it had developed one-fifth of Limbourg (Flanders) to establish a colony of 2,500 paupers and 4,000 orphans there.[62] Giving land to the poor would attach them to property, suppress begging, and ensure civil peace: farewell, wrote Rauch, to 'the multitude of idlers, the disturbing, factious rabble that was ever the hope and tool of the parties'.[63]

That Rauch, despite his tendency to promise too much, was in tune with the realities of his time is confirmed by the fate of his fourth and last company, in a manner at once ironic and cruel. In May 1828, he founded the General Drainage Company, then managed to surround himself with prominent figures who acted as advisers and conferred credibility on the undertaking: scientists such as Fourier, Tessier and Mirbel, along with senior officials at the Ponts such as Prony and Girard.[64] The principle of this company was to work in synergy with the state. First, it called on the latter to take control of lands belonging to third parties (private individuals or communes), in the name of a developmental imperative. This being done (often by expropriation), the state granted these parcels to the company. The latter then paid the costs of drainage prior to returning the lands, while keeping some of them. It made its profit by re-selling what was now agricultural land.

Operations were launched in fourteen departments: Camargue, Aube, Marne, Cher, Gard, and so on. In 1833 the company's capital was nearly

60 For example, *Annales européennes*, vol. 2, p. 47 and vol. 7, p. 257.

61 Ibid., vol. 10, p. 189.

62 Ibid., vol. 8, p. 188.

63 Ibid., vol. 9, p. 274.

64 AN F14/11167 and Jean-Paul Haghe, *Les Eaux courantes et l'État en France (1789–1919)*, geography thesis, EHESS, 1998, pp. 83–6.

2 million francs.[65] Of course, it was not all plain sailing: some proprietors resisted expropriation by doing the drainage themselves, or challenging the procedure, or physically preventing the works.[66] Furthermore, the legislative reform proposal presented in the Chamber at the same time by the financier Jacques Lafitte (the Company's banker[67]), intended to facilitate the transfer of land, was voted down. Despite everything, the company managed to retain the confidence of its shareholders and to operate until the mid-1840s.

A certain success, then, but a cruel one for Rauch, for he had in the interim been forced to resign as manager and leave the company he had co-founded. He died in 1837 in a bedroom that had been lent to him, with a few personal belongings and a list of debts – those of a man just scraping by.[68] His crusade lay in the past: he devoted his last years to an enterprise of forced-march development, focused on cultivation and resale, far removed from the grand visions of the botanical, climatic and moral restoration of France that had once inspired him.

65 *Compagnie générale de dessèchement. Assemblée des actionnaires du 27 avril 1833*, Paris: Félix Locquin, 1833.

66 Haghe, *Les Eaux courantes*, pp. 83–6.

67 At least until the collapse of his banking activities at the start of the 1830s.

68 Inventory following the death of François-Antoine Rauch, act of 12 April 1837 (AN ET/III/1498).

11

Circular n° 18

An Inquiry into Climate Change from Two Centuries Ago

It is one of those finds that surprises the historian and adds momentum to an investigation – sometimes culminating in a book. In our case, it took the form of one and then another box stuffed with documents, unearthed from the archives of Météo France and the Académie des sciences de Paris.[1] Inside were several thousand pages of letters, notes, reports, and lists. Some numerical data (very little). Some graphs, but no pictures. Multiple signatures: those of prefects, sub-prefects, mayors, farmers, naturalists, engineers, rentiers, and local men of letters. Some precise, concise, circumspect responses; others that are confused, tortuous and occasionally exhaustingly repetitive. All are voices which, at a distance of two centuries, speak to us of climate change in France, its causes and effects, and many other things besides.

These are the responses to Circular n° 18, sent out at the end of April 1821 by the 'sciences and fine arts' office of the French Interior Ministry. Addressed to all prefects, the circular began with an observation: the growing prevalence over several years of 'palpable cooling of the

1 Méteo France's box has already attracted the attention of Vincent Bainville and Philippe Ladoy, 'Préoccupations environnementales au début du XIXe siècle. La circulaire no 18 du 25 avril 1821', *La Météorologie*, April 1995, pp. 88–94.

atmosphere', 'sudden variations in the seasons', and 'extraordinary storms or floods'.[2] What was going on? And, if it existed, was this trend due to the deforestation of France? This is what the inquiry sought to determine, posing five questions about the development of departments' 'meteorological system' (wind, rain, snow, but also floods, rivers and glaciers), and of their forests (surface area, composition, ownership), in the last thirty years – that is, *since the Revolution*. In the early nineteenth century, for the first time, a European state was launching a national inquiry into climate change and man's possible responsibility for this ongoing process.

The Ministry of the Interior and of the Climate

Why this initiative? What were its origins, goals, and ulterior motives? To understand this, we must go back two years, to the beginning of 1819, when a small group of agronomists and naturalists, all of them convinced of the seriousness of climate change, found themselves propelled to positions of influence in the Interior Ministry – the heart of government in France.

On 18 January 1819, the minister and head of government Élie Decazes set up an Agricultural Council. The idea was to promote a more modern, more rational agriculture, thanks to a body of expertise integrated into the ministry. It comprised notable scientists, recruited in the main from the Académie des sciences and the Société royale d'agriculture, including Jean-Antoine Chaptal, a chemist and major industrialist, who had been interior minister under Napoleon. The Council was assisted by provincial correspondents who were to devote part of their land to agronomic experiments prescribed from Paris.

The party of climate change found itself in a position of power in the Council. Why? Because the latter was closely linked to the activities of a botanist, Charles-François Brisseau de Mirbel.[3] Mirbel was among the scientists who most vigorously sounded the alarm about climate change. In the botany course he delivered at the Sorbonne, he warned that if France did not protect its trees, it would suffer the same fate as the lost civilizations

2 'Circulaire du 25 avril 1821. Intérieur. Sciences et Beaux-Arts', no. 18', Météo France collection.

3 Anselme Payen, 'Éloge historique de M. de Mirbel', Paris: Vve Bouchard-Huzard, 1858; Élie Margollé, *Vie et travaux de M. de Mirbel*, extract from *Revue germanique et française*, Saint-Germain: L. Toinon et Cie, 1863.

of Egypt and Judea.[4] Mirbel was also a friend of Minister Decazes, whom he pressed to create the Agricultural Council. The two men had become friends when both were advisers to the King of Holland, Louis Bonaparte, in 1806–08. In September 1815, when Decazes became, first, the minister of Police, Mirbel joined him as secretary general of the ministry. And once Decazes had moved to the Interior Ministry (in December 1818), Mirbel occupied the same key post there. It was he who decided the appointments to the Agricultural Council, recruiting his closest colleagues. Thus, he brought in André Thouin, who as early as 1784 had adopted the climatic theories of Pierre Poivre and Bernardin, and Louis Bosc d'Antic, another naturalist long persuaded of the dangers of anthropogenic climate change.[5]

In January 1820, a cold spell hit France. In Paris, the Seine froze over and people died of cold in the streets.[6] In south-eastern France, arboriculture took a blow: trees were killed off by frost or had to be cut down. 'The orange trees of Hyères and Nice', wrote the *Journal des débats* on 2 February, 'are entirely lost . . . and the olive trees have been very badly damaged in much of Provence.'[7] The *Journal de Marseille* went one better: 'Fig trees are in the same condition as the olive trees. Wherever they were already pruned, the vines have suffered greatly.'[8] Most ominous was the damage to olive groves. Because it took nearly twenty-five years to obtain a productive tree, this augured a tremendous economic loss.[9] The very existence of the olive oil industry in France seemed threatened.

The Agricultural Council was urgently mobilized in response. Louis Bosc d'Antic was tasked with investigating the death of olive trees.[10] He questioned his Provençal correspondents, who assured him that climate deterioration was indeed the cause of the disaster.[11] Since 1816 and 'the

---

4 Charles-François Brisseau de Mirbel, Éléments de physiologie végétale et de botanique, Paris: Magimel, 1815, vol. 1, pp. 447–53.

5 'Climat', *Nouveau Cours complet d'agriculture*, Paris: Déterville, vol. 4, 1809, pp. 113–19. Bosc was also a subscriber to Buffon's cooling theory, to the point of advising against the reforestation of mountains, which were doomed to be frozen in ice.

6 *Journal des débats*, 6, 9, 11, 12, 13, 17, 18, 19, 20 and 31 December 1820.

7 Ibid., 2 February 1820.

8 *Journal de Marseille*, 5 February 1820.

9 Jean Nicod, 'Grandeur et décadence de l'oléiculture provençale', *Revue de géographie alpine*, vol. 44, no. 2, 1956, pp. 247–95.

10 *Rapport sur les travaux du Conseil d'agriculture, pendant l'année 1820*, Paris: Imprimerie royale, 1821, p. 19. Archives nationales, F10/275.

11 *Collection de mémoires ou de lettres relatives aux effets, sur les oliviers, de la gelée du 11 au 12 janvier 1820*, Paris: Mme Huzard, 1822, pp. 17–19, 139–40.

year without a summer', the Provençal elites had been convinced that climate change was underway and threatened the whole region with ruin.[12]

Others, however, reckoned that it was necessary to hold firm and reassure farmers, in order to stop them giving up on olive tree cultivation at the very point – after their recent losses – when they had to decide what to replant. This was the objective of Adrien de Gasparin, an agronomist from the Vaucluse (who would become interior minister under the July Monarchy). All his efforts went into persuading the farmers of Provence 'that the seasons have a *regular, permanent course*, dependent on the general laws of the universe, and consequently *immutable like it*, and that their greater or lesser variations are simply oscillations around a fixed point from which they cannot greatly stray'.[13] For Gasparin, something vital was at stake: saving southern oleiculture.

The winter of 1819–20 marked the high point of anxieties about deterioration of the French climate. It represented the climax of a sequence that began at the end of the previous century and was accentuated by the explosion of Tambora. At the start of 1821 the Agricultural Council decided to settle the question: after years of debates, the state must take things in hand and mobilize all its resources to conduct an inquiry. Minister Joseph Jérôme Siméon, who had succeeded Decazes in February 1820, gave the green light.[14] Circular n° 18 was issued.

From the government's point of view, the initiative had several objectives. Firstly, the inquiry responded to the popular and media commotion created by the freezing temperatures of 1819–20. Here was a way for the regime to display its determination to act on a problem on which it had, in reality, no immediate purchase.

At a time when sales of national forests were accelerating, requiring an annual vote in Parliament, it was also a way of countering attacks by ultra-royalists. For the inquiry was not without ulterior motives. It was not simply a question of studying climate change, but also of steering the search for culprits, with a view to blaming, not the recent sales, but the disastrous consequences of the Revolution. The circular's request – to

12 On 22 February 1817 the Var deputy Jacques Aurran-Pierrefeu thus reported to the Chamber, opposing a new tax on oil production.

13 Adrien de Gasparin, 'Mémoire sur la culture de l'olivier dans le midi de la France', *Bibliothèque universelle de Genève*, vol. 7, March 1822, pp. 49–74 (here, p. 71).

14 *Procès-verbaux de l'Académie des sciences de Paris*, vol. 8, 1824–27, p. 26.

examine forestry and climate development over the last thirty years, that is, since 1789 – was clear enough on this point. To formulate the question in these terms had the advantage of mollifying the ultras, while exonerating the Restoration's recent forestry policy.

But that was not all: the inquiry also sought to demonstrate the crucial climatic role of *mountain* forests, which were *not involved in the sales decided since 1814*. The manoeuvre was subtle. Rather than denying the existence of climate change outright, it pointed the finger at different culprits: the mountain dwellers who destroyed their own environment and harmed the whole country's climate.

This charge sprang from the conviction, widely held at the time, that French mountains had been damaged.[15] The guilty parties were local people and their archaic agro-pastoral practices. Mirbel was sure of it: he had been trained in botany by Ramond de Carbonnières, who was one of the first to theorize and popularize this notion in the late eighteenth century.[16] For four years under the Directory, when he had had to flee Paris for political reasons, Mirbel explored the Pyrenees in Ramond's company and it was these excursions that shaped his declinist view of mountain realities. Two decades later, now at the Interior Ministry, he appointed his former mentor to the Agricultural Council. Mirbel had remembered his lessons well. In August 1820, reacting to the freezing of Provençal olive trees, he promoted in the press the *Project de boisement des Basses-Alpes* published by a former prefect, Pierre-Henri Dugied, the previous year.[17] For Mirbel, this memoir was a boon: not only did it denounce the injury to the mountains, it also stressed the specifically climatic *effects* of this. The deforestation of the Alps, Dugied explained, affected the climates of Provence, Burgundy and the Franche-Comté. Had it not been seen, he

15 See, in particular, Jean-Paul Métaillé, 'Lutter contre l'érosion: le reboisement des montagnes', in Andrée Corvol, ed., *Les Sources de l'histoire de l'environnement. Le XIXe siècle*, Paris: L'Harmattan, 1999, pp. 97–106.

16 As regards the Pyrenees, see, for example, Louis Ramond de Carbonnières, *Lettres de M. William Coxe à M.W. Melmoth sur l'état politique, civil et naturel de la Suisse*, Paris: Belin, 1781, pp. 192, 276 (translation and additions by Ramond); Serge Briffaud, 'Le rôle des catastrophes naturelles. Cas des Pyrénées centrales', in Andreé Corvol, ed., *La Nature en Révolution. 1750–1800*, Paris: L'Harmattan, 1993, pp. 134–44.

17 Charles-François Brisseau de Mirbel, 'Des forêts', *Journal des maires*, 15 August 1820. Pierre-Henri Dugied, *Projet de boisement des Basses-Alpes*, Paris: Imprimerie royale, 1819. Dugied had initially composed it as a handwritten memo addressed to the Agricultural Council. *Rapport sur les travaux du Conseil d'Agriculture pendant l'année 1819*, Paris: Imprimerie royale, 1820, pp. 15–17. Archives nationales, F10/275.

asked, *prior* to the winter of 1819, that the olive trees of Provence 'have been freezing more often than before over the last twenty-five to thirty years'?[18] Here the nature of the threat changed: at stake was not only the integrity of the mountain environment, but the risk of climatic disruption to the major agricultural regions, where wealth and population were concentrated.

This account also had a major political advantage. Mountain-dwellers were the ideal culprits: for thirty years their agro-pastoral practices – the use of common pastureland in particular – had been accused of hastening deforestation and causing soil erosion and floods. To render them responsible for the cold weather of 1820, in the same way as the revolutionaries and peasants of Year II, made it possible to exonerate the policies of the government to which Mirbel and his Agricultural Council belonged.

Deciphering Change

The inquiry questionnaire was sent out to prefects at the end of April 1821, and in mid-June the minister of the interior informed the Académie des sciences, instructing it to summarize the forthcoming responses.[19] As they came in, these were transmitted to an academic commission comprising the agronomist Victor Yvart and two members of the Agricultural Council, Jean-Baptiste Huzard and Louis Bosc d'Antic. This was not the first time that French climates had been the subject of state statistics. The Royal Societies of Medicine and Agriculture, by establishing networks of meteorological observation in the 1770s and 1780s, had already sought to supply the royal power of which they were extensions with tools for managing illnesses and harvests.[20] In the 1800s, the launch of a major survey, Departmental Statistics of France (DSF), aimed to draw up a complete anatomy of post-revolutionary France. And because its interpretative grid

18 Dugied, *Projet de boisement des Basses-Alpes*, p. 7.

19 *Procès-verbaux de l'Académie des sciences de Paris*, vol. 7, 1820–23, p. 205. On the inquiries of the Restoration (but on the state of public opinion), cf. Pierre Karila-Cohen *L'État des esprits. L'invention de l'enquête politique en France, 1814–1848*, Rennes: Presses universitaires de Rennes, coll. 'Carnot', 2008.

20 Louis Cotte (cf. Chapter 4) was the project manager of the network of the Society of Medicine, under the auspices of the doctor Félix Vicq-d'Azyr. On these observation networks, see in particular Jean-Paul Desaive et al., *Médecins, climat et épidémies à la fin du XVIIIe siècle* and Fabien Locher, 'Le nombre et le temps. La météorologie en France (1830–1880)', doctoral thesis, EHESS 2004, chapter 1.

was a view of the country conceived in neo-Hippocratic terms, it solicited from participants topographies of their habitats in order to compare the climates of departments and the 'nature' of local inhabitants.[21]

The 1821 inquiry was unique in being wholly devoted to climate change. The responses it elicited are today preserved in two archival collections, one held by the Académie des sciences, the other by Météo France. These last documents were probably bequeathed by the Ministry of the Interior, at the end of the nineteenth century to one of the bodies of which Météo-France is the heir – probably the Central Meteorological Office (1878–1921). The itinerary of the material is, however, somewhat mysterious. Indeed, fifteen years after the start of the inquiry these responses were thought to have been lost when, in 1838, an unsuccessful attempt was made to find them (Chapter 12). Had they been mislaid before re-surfacing and then being handed over, decades later, to the state's meteorological officials? The bundles preserved in the Académie had a calmer history: they are copies of the responses to the inquiry, progressively passed on by the Ministry to feed into the academicians' synthesis. In all, we possess responses from sixty out of eighty-six departments (fifty-five at Météo France and fifty at the Académie). Inventories drawn up at the time confirm that this is the complete corpus.[22] A little under one-third of prefects never replied to the questions of Circular n° 18.

At the time, the inquiry had a certain exposure in the media. Rauch loudly congratulated himself in his *Annales* for having prompted a measure that was, he wrote, 'perhaps the most crucial that has ever emerged from the ministry'.[23] He reproduced the questionnaire and a (not unbiased) selection of responses from thirty departments.[24] Having obtained

21 On the DSF, see Marie-Noëlle Bourguet's book *Déchiffrer la France. La statistique départementale à l'époque napoléonienne*, Paris: Édition des Archives contemporaines, 1989. See also Alain Desrosières, *La Politique des grands nombres. Histoire de la raison statistique* ([1993], Paris: La Découverte, 2000; Jean-Claude Perrot, *L'Âge d'or de la statistique régionale française*, Paris: Société des études robespierristes, 1977. One of the DSF's side issues concerned possible climate changes. Chaptal's circular of 19 Germinal Year IX to prefects: 'One might research whether the air temperature has changed, and why; whether land clearances have influenced it.' Document reproduced in Bourguet, *Déchiffrer la France*, p. 419.

22 See the partial lists preserved in the Académie and Météo France collections and the *Procès-verbaux de l'Académie des sciences de Paris*, vol. 8, 1824–27, p. 27.

23 *Annales européennes*, vol. 3, 1821, p. 127.

24 Ibid., pp. 15–19 and pp. 392–402 (Var). See also *Mémorial de l'industrie française, des sciences et des arts*, vol. 5, April 1821, pp. 201–04.

permission to receive copies on arrival, he milked this privilege to cast himself, in exaggerated fashion, as the instigator of the undertaking and its virtual project manager.[25] For his part, Rougier de la Bergerie grumbled in his *Cours d'agriculture pratique* (which was actually a journal) about the favouritism of the minister, who preferred Rauch to him.[26]

The responses form a kaleidoscope of observations, comments and queries from the whole country. The questionnaire was sent on from the head of the departmental executive to a range of other interlocutors: to sub-prefects and then to mayors and agrarian property-owners; to the ad hoc committees sometimes set up by prefects; to local scientific societies that provided expertise and sometimes published the results in their journals.

Some apologized in advance: 'I have only received some superficial, contradictory and purely hypothetical responses', wrote the prefect of Haute-Marne candidly. 'The summary I shall make of this material will therefore afford no enlightenment.'[27] Others thought it was a matter for scientists: 'Many skilful naturalists and physicists have dealt with this question. . . . If such distinguished scientists cannot agree as to the influence of the shelter procured by forests . . . I shall not venture to enter into the discussion of a subject I know so little about'[28] (prefect of the Ardennes). Several used an a priori argument to avoid responding: since their department had not suffered deforestation, they had nothing to say about climate change. Finally, some did not mince their words: 'These mad requests must stop', snapped one correspondent after the umpteenth reminder from Paris.[29]

However, most prefects provided substantial responses, usually summarizing the documents produced by their local interlocutors, which were appended. No unequivocal diagnosis of changing climates in France emerges from this mass of correspondence and documentation. Several key elements are prominent nevertheless, over and above the diversity of perceptions, assessments, and hypotheses.

25 François-Antoine Rauch, 'Demandes à faire à monsieur Grille', n.d., and missive to M. Blanchard, 27 September 1822, Météo France collection.

26 *Cours d'agriculture pratique*, vol. 7, Paris: Audot, 1822, pp. 49–51.

27 Letter of 12 November 1821 from the prefect of Haute-Marne to the interior minister. 'Haute-Marne' file, Météo France collection.

28 Undated letter from the prefect of the Ardennes to the interior minister, 'Ardennes' file, Météo France collection.

29 Bainville and Ladoy, 'Préoccupations environnementales au début du XIXe siècle', p. 91.

First of all, one overwhelming consensus: the climate had become *more variable, more erratic*. Since the Revolution, the seasons had been less regular, more changeable in intensity, and their boundaries fluctuated: 'It is indisputable that the seasons no longer have the same regularity, and this difference has been palpable for fifty years' (sub-prefect of Tournon);[30] 'a clearly recognized variability in the times of the periodic return of the seasons' (prefect of Bouches-du-Rhône); 'less clear-cut seasons' (general secretary of the prefecture of Ille-et-Vilaine).

This increasing irregularity of the seasons was matched by increasingly erratic daily or weekly 'weather conditions'. Now, it was everywhere reported, wind, rain, cold and heat arrived more abruptly, in bursts: 'the temperature is infinitely more variable than formerly' (adviser to the prefecture of Gers); 'an unquestionable . . . disruption of the temperature' (prefect of the Basses-Alpes); more changeable winds (prefect of the Hérault).

Climate and atmospheric conditions were thus said to be increasingly variable. But what of climatic *averages*, and especially temperatures – the threat of a cooling of France having been key in the launching of the inquiry?[31] This was one of the major lessons of the responses, but it took the form of a negative result. In fact, there was no consensus on this crucial point. Certainly, the idea of a change was widespread. In a minority were those who, like the prefect of the Hérault, referred to a constant heat, writing that 'it has been observed for sixty years that [harvests] of grain, wine and fodder are ready at the same time and scarcely vary by a few days, earlier or later, save in particular circumstances.'[32] Or like the mayor who derided, for their belief in climate change, 'those men who believe that nature proceeds as rapidly as their speculative mind'.[33] The best-argued scepticism came from the Count of Montlosier, a naturalist called on to discuss his own department, the Puy-de-Dôme.[34] For him the 'summits of

30 *Annales européennes*, vol. 3, 1821, p. 385. He had changed his mind since his first response to the ministry ('Ardèche' file, Météo France collection).

31 There are two separate questions here, for a magnitude changing over time may well become more variable around its average value without the latter changing, and vice versa.

32 Letter of 9 May 1822 from the prefect of the Hérault to the interior minister. 'Hérault' file, Météo France collection.

33 Letter from Mareschal, mayor of Meyreuil. 'Bouches-du-Rhône' file, Météo France collection.

34 *Annales européennes*, vol. 6, 1824, pp. 386–96.

the sky' contained such large masses of water and ice that human action was probably negligible. Meteorology was to be explained exclusively by meteorology: 'The study of these phenomena pertains to the study of the atmosphere, and has no affinity with the petty action of our deserts or our forests'.[35] For him, there was no sign of climate change, only masses of atmospheric air indifferent to the existence of men, which produced seasons and celestial phenomena.

Many responses claimed that there had been a change in temperatures. But in which direction? For some, winters were becoming milder: 'Previously long and heavy winters are today little more than lovely autumn days' (prefect of the Ariège); 'winters are less cold' (Gallucheau, for the Saintes Agricultural Society). For others, by contrast, there was no doubt that it was colder, or cold for longer. 'It seems certain, according to the testimony of the oldest inhabitants, that the temperature has fallen in the last thirty years' (sub-prefect of Parthenay); 'the cold is no longer as sharp, but it lasts longer' (sub-prefect of Bressuire).[36]

Nor was there any consensus over other climatic phenomena. Thus, as regards winds: 'The winds are generally stronger and colder' (prefect of the Var); 'the winds have been less violent' (prefect of the Vendée); 'the winds have not been more violent, or more destructive, or more variable' (prefect of Seine-et-Oise).[37] Now – and this is an important point – if the responses seem contradictory and chaotic when surveyed en masse, they hardly acquire greater coherence when broken down by geographical zone, region, or type of terrain. Only the Provençal departments, still traumatized by the winter of 1819-20, unanimously complained of climate cooling.

However, one reference largely structured the responses to the inquiry: the *water cycle* as explanatory schema, cause for concern, and motive for injunctions to act. If deforestation had made its effects felt since the Revolution, it was primarily by rendering the water cycle more erratic. This consideration broadened the idea of a more variable climate as regards precipitations, extending it to springs, fountains, river flows. In

35 Ibid., p. 389.

36 'Ariège', 'Charente-Inférieure' and 'Deux-Sèvres' files, Académie des sciences collection. Overall, one assessment was strangely widespread but difficult to explain: a reduction in snowfall since the Revolution (Gers, Basses-Alpes, Mayenne, Haute-Marne, Meurthe, Maine-et-Loire, Hautes-Pyrénées, Aveyron, Vendée, etc.).

37 'Var', 'Vendée', 'Seine-et-Oise' files, Météo France collection.

the responses, two major schemas were employed to explain the great disruptions observed from clouds to riverbeds.

The first schema was centred on the *pedological* action of trees. Forests were said to act by stabilizing the soil, encouraging the build-up of humus, mixing their roots with the earth. Water soaked into this upper layer before feeding underground reservoirs (and therefore springs) and flowing smoothly over the surface. But in the absence of trees, rainfall was no longer absorbed and water no longer retained in the soil or subsoil. Hence stop-start flows, dry springs, floods. Above all, the responses censured what was happening in the mountains, where with fewer trees the soil itself was swept away by the rains.

The second schema was centred on the *meteorological* action of trees. Once thinned out or destroyed, forests could no longer fulfil their role as attractors of clouds and hence as steady 'pumps' of atmospheric water. This meant more irregular precipitations and thus torrential rain, droughts, the erosion of slopes, and flooding. Louis-Augustin d'Hombres-Firmas, correspondent of the Académie des sciences in the Var, was one of those who investigated this idea in most depth, revealing in passing all he owed to Bernardin de Saint-Pierre and Rauch: the trees that used to serve as 'electric conductors' had been cut down so they no longer attracted clouds, no longer acted as conduits transferring the clouds' water into the soil.[38] Others rounded out the picture with details that were likewise traceable to Rauch. Thus Dralet, a forest keeper writing from Toulouse, mentioned the risks posed by the freezing at altitude of the volumes of water that were no longer attracted by trees – an argument inspired by the *Régénération de la nature végétale*. A prolific author, a pioneer of conservation of the Pyrenees mountains, Dralet was a veteran of mountain hikes in the company of Ramond and Mirbel.[39] Sometimes the references overlapped, as with an anonymous informant from the Pyrénées-Orientales who combined the theories of Bernardin and Rauch with Buffon's fire at the Earth's core.

38 Louis-Augustin d'Hombres-Firmas, 'Essai sur le déboisement des montagnes en France', n.d. 'Gard' file, Météo France collection.

39 See also the response by Parison, professor of physics at the College of Épinal and responsible for expert assessment of the Vosges. Like Rauch, he spoke of trees as siphons and of the sinister perspective of a France transformed in the image of the great deserts of Africa and Asia. 'Haute-Garonne' file, Météo France collection, and 'Vosges' file, Académie des sciences collection.

Pointers, Evidence and Testimony

The empirical foundation of the responses was weak. Many items of correspondence regretted the lack of numerical meteorological data: 'The absence of meteorological observations at the time when the forests were still intact does not [allow] us to make exact comparisons' (prefect of the Jura).[40] 'There isn't', deplored the prefect of the Aube, 'throughout the whole of the department . . . either in the teaching body or outside it, a single physicist or naturalist, in a word, anyone who has been concerned to note the variations of seasons, the direction of the winds.'[41]

Some scattered data were nonetheless available. This was true, for example, of the Bas-Rhin, where Herrenschneider, physics professor in the Strasbourg faculty, used his own observations to argue for the constancy of climates.[42] Faced with the scarcity of observations, proposals for initiatives re-emerged. Thus the Literature, Sciences and Arts Society of Rochefort considered 'that it will only be possible to resolve the issue . . . when the government has made daily meteorological observations obligatory, at least in each principal town or capital of the sub-prefecture'.[43] This proposal actually took up a suggestion made by Rauch in his successive books, of creating a network of such meteorological and 'hydro-vegetal' observatories throughout the country. Such was also the spirit of the intervention by Girou de Buzareingues, a property-owner in the Aveyron and correspondent of the Agricultural Council, who put forward nothing less than a network of meteorological observatories on all of France's coasts, mountains and borders, and even (modestly) 'in all the states of the world'.[44]

Meanwhile, the inquiry took place at a time – the first third of the nineteenth century – when meteorological observation using instruments was very rare. By and large, only in astronomical observatories,[45] and in the

40 Letter of 9 May 1822 from the Prefect of the Jura to the interior minister. 'Jura' file, Météo France collection.

41 Letter of 3 April 1822 from the Prefect of Aube to the interior minister. 'Aube' file, Météo France collection.

42 In some instances, old data was dug up. Cf., e.g., the 'Ariège' file, Académie des sciences collection.

43 Copy of a letter from the society to the prefect of Charente-Inférieure, n.d., 'Charente-Inférieure' file, Météo France collection.

44 Proposal from Girou de Buzareingues, sent by the general secretary of the Aveyron to Baron Cuvier, 27 October 1821. 'Aveyron' file, Académie des Sciences collection.

45 Among the rare data discussed were measurements of temperature carried out at the Paris Observatory in the 1800s and 1810s. A.J.B.L. Doulcet, *Mémoire sur la destruction*

homes of some amateur practitioners, were 'weather conditions' recorded on a daily basis. This was a trough between the late-eighteenth-century golden age of medico- and agro-meteorological networks and the revival of observation fostered by the new, so-called 'dynamic' meteorology, from the 1830s and 40s.[46] The 1821 inquiry confirmed the quasi-disappearance of the practice of logging the weather that was so common at the end of the ancien régime and would subsequently return. Add to this an important feature of the inquiry: forestry managers, who were its preferred interlocutors, were still mostly recruited at the time on the basis of a training in law – the Nancy forestry school was only founded in 1824 – and their lack of familiarity with the culture of accurate observation was patent.[47]

No less striking about the responses to the inquiry is the process of memorial and institutional erasure, which meant that the enormous mass of data collected by the Société royale de médecine before the Revolution was quite simply ignored – despite the huge wealth of information it contained, including in terms of numerical climatic data, as shown by the collective investigation carried out in the early 1970s by Emmanuel Le Roy Ladurie, Jean-Pierre Goubert and Jean-Pierre Peter in particular.[48] This erasure was enduring: twenty years after Circular n° 18, when the General Statistics of France attempted to produce a climatological description of the country by collecting data, this mine of information was still not used or even mentioned.[49] What Jean Meyer once called 'the most important of all the administrative inquiries of the eighteenth century' would lie forgotten in the archives for almost two centuries.[50]

Without any data to inform responses to the inquiry, recourse was frequently had to human memory, especially that of 'old people', who were questioned but of whom it is hard to say – because they are never named – whether they are much more than a source of discursive authority.

des forêts, sur les effets qui en résultent, et sur les moyens de retarder et de réparer leurs pertes, Auxerre: Imprimerie Le Coq, 1821, pp. 13–14.

46 Locher, *Le Savant et la Tempête*, pp. 13–27, 63–79.

47 On these foresters, some of whom had returned from exile, see Andrée Corvol, *L'Homme aux bois. Histoire des relations de l'homme et de la forêt (XVIIe–XXe siècle)*, Paris: Fayard, 1987, p. 245.

48 Jean-Pierre Peter, 'Une enquête de la Société royale de médecine sur les épidémies, 1774–1794', *Annales ESC*, vol. 22, no. 4, 1967, pp. 711–51; Desaive et al., *Médecins, climat et épidémies à la fin du XVIIIe siècle*.

49 Fabien Locher, 'Le rentier et le baromètre: météorologie savante et météorologie profane au XIXe siècle', *Ethnologie française*, no. 39, 2009, pp. 645–55.

50 Desaive et al., *Médecins, climat et épidémies à la fin du XVIIIe siècle*, p. 9.

Votive festivals occupy a big place in such testimonies. For example, in Corsica where, in the villages of Celano and Orcino, it was customary to celebrate 1 May with cakes baked on the new leaves of chestnut trees. These days, however, to the great surprise of the 'old people', no leaves were to be found.[51] Or in the Var, where the same old people 'say that they regularly had abundant rain in the last fortnight of September or the first fortnight of October, which they called Saint Michael's rain', but which was no longer seen today.[52] In this recourse to the testimony of old folk, the report of whatever they had to say was open to challenge almost in the same breath: was not their perception clouded by the 'natural tendency' older people have always had to 'represent past times as happier in every respect' (prefecture of Gers)?[53]

Without quantified observation, and conscious of the fragility of memories, prefects turned to a third type of source: proxies. Above all, traces, material or written, of lost agricultural cultures. We have seen how, in the late eighteenth century, meteorologists intent on being 'historiographer[s] of nature' (Chapter 4) had already turned to these when faced with the unreliability of observations, standards and old instruments. Prefects and their informers did likewise. In the Ardèche, according to a sub-prefect, deeds from 1634 referred to plots as 'small' and 'big olive grove', where this culture was impossible today; land registers recorded 'paths between the vine rows' at heights where they could no longer be grown.[54] In the Cantal, 'the old land registries of the cathedral chapter of Saint-Flour prove that in the thirteenth and fourteenth centuries one levied the tithe on the harvest of vines positioned on the southerly slope of the mountain.' These documents, concluded the naturalist Devèze de Chabriol, 'are proof that this kind of culture was pushed back by the climate to the lower . . . parts' of the department.[55] The history of the retreat of crops to lower latitudes or altitudes served to back up a diagnosis of climate change otherwise impossible to characterize. Sometime eyes even turned to . . . roofs – as in

51 'Note sur les causes qui ont pu amener des refroidissements sensibles', n.d., 'Corse' file, Météo France collection.

52 'Réponses aux questions proposées par son excellence le ministre de l'Intérieur', 14 March 1823. 'Var' file, Météo France collection.

53 Letter of 18 November 1821 from the adviser to the Gers Prefecture to the interior minister. 'Gers' file, Académie des sciences collection.

54 Letter from the sub-Prefect of Tournon to Rauch, published in *Annales européennes*, vol. 3, 1821, pp. 383–92 (here, p. 391).

55 *Annales européennes*, vol. 9, 1825, pp. 395–6.

the Perche, where old buildings had beams of chestnut wood, with chestnut trees, whose disappearance (the prefect of Eure-et-Loir explained) was due to the heavy winters of 1709 and 1740.[56]

An even more original endeavour, with a fine future ahead of it (cf. Chapter 13), hailed from the Yonne.[57] The head engineer of the Ponts et chaussées departmental service, Jean-Baptiste Simon Fèvre, suggested using *bans de vendanges* (harvest benches) in Burgundy – which indicated the start of the grape harvest – for pointers about the climate. Thus, he argued, one could go back centuries, and he appended to his response a form to be filled in to this end. But, for now, the idea remained a dead letter.

Another way consisted in using the levels of waterways to identify past climate modifications, as had been done in the eighteenth century. The Baron de Malaret, president of the Agricultural Society of the Haute-Garonne, presented the results it had obtained by consulting observations of river levels, courtesy of a local scholar and the administration of the Midi canal.[58] The society thus had access to a series dating back to before the Revolution, from which it derived graphic representations in the form of curves. Yet it was difficult to reach a conclusion: the flooding of the Garonne appeared greater in recent times; elsewhere, however, physical reference-points marked on the river's banks indicated record highs *before* the Revolution. The memory of the hydrological past, too, seemed hazy.

Scales of Change

Reading through the responses, one thing is clear: in the early nineteenth century, the climate and climate change were by no means regarded as purely local in their features, manifestations, development, and causes. In other words, it was certainly not the consideration of large spatial scales that distinguished ways of thinking about climate change in this era from those of our time. There is no room here for some convenient

56 Letter of 25 May 1822 from the Prefect of Eure-et-Loir to the interior minister. 'Eure-et-Loir' file, Académie des Sciences collection. We also find this reference to the chestnut trees' former presence, visible in old beams, in the responses from Lot and Loiret. Like olive trees, vines and orange trees, chestnut trees were prime indicators of climate.

57 'Yonne' file, Académie des Sciences collection.

58 Report by the Baron de Malaret, dated 28 May 1822. 'Haute-Garonne' file, Académie des sciences collection.

contrast between a climate change perceived and thought of as local (as in the nineteenth-century) and perceived and thought of as global (as in our time).

This does not mean that no one thought of climate change as a process whose causes and effects were spatially highly circumscribed. But there was no consensus on this point: everyone pushed their own conception of the spatial scale of change – a conception that was often bound up with pointing to a cause or a culprit for it.

For some, the planet itself was the appropriate scale for grasping the climate question. Thus, it was the 'well-nigh total destruction of forests throughout Europe and the New World' which, according to the sub-prefect of Privas, explained the climatic disruption observed in the Ardèche.[59] 'Clearances in northern Germany and North America' were said to have affected the Landes.[60] In his report to the prefect of Loir-et-Cher, Guérin d'Ogonière from the department's agricultural society spoke of the Earth stripped of 'at least half of its forests', including 900 million acres lost in Europe and 98 million in France.[61] 'So is it any wonder', he continued, 'that such immense empty spaces at all points of the globe have changed the direction of terrestrial winds?'

On a smaller scale, it was the deforestation of French mountains, coastal departments (which suffered from winds off the sea), and neighbouring departments that explained current developments. 'It is not so much . . . in this department', argued the prefect of Pas-de-Calais, 'as in adjoining ones, where major operations are being carried out, that we shall have to look for the causes whose effects are felt here.'[62] The local country, wrote his neighbour from the Somme, is 'influenced by operations carried out far away in the mountainous and forested departments.'[63] This kind of argument was also handy, offloading potential responsibilities onto other authorities, economic actors or public servants.

59 'Rapport – Ardèche', 10 December 1821 ('arrondissement de Privas' sub-section). 'Ardèche' file, Académie des sciences collection.

60 'Variations atmosphériques. Département des Landes', report of 19 November 1821. 'Landes' file, Météo France collection.

61 Letter of 19 May 1822 from Guérin d'Ogonière to the prefect of Loir-et-Cher. 'Loir-et-Cher' file, Météo France collection.

62 Letter of 26 February 1822 from the prefect of Pas-de-Calais to the interior minister. 'Pas-de-Calais' file, Météo France collection.

63 'Observations générales' dated 18 June 1822. 'Somme' file, Météo France collection.

Another lesson of the inquiry was the persistence of conceptions that inscribed climatic phenomena in continuums embracing telluric and/or astronomical phenomena – hence causal chains of far-reaching scope, expressed on the occasion of particular events, often helping to present an alternative to explanations involving deforestation. Earthquakes, especially the one in Calabria in 1783, were blamed. The changes 'seem to have to be attributed not so much to particular deforestations . . . as to a large number of physical causes, among which is the earthquake in Calabria' (Eure-et-Loir);[64] some 'thought they had noticed that the changes that have occurred in the condition of the atmosphere have become more tangible since the earthquakes of Lisbon and Calabria in 1755 and 1783' (Charente).[65]

Other responses sounded like an echo of the post-Tambora debates, by alluding to the North Pole. Thus the sub-prefect of the Gers, who attributed the current cooling to a great collapse of ice that had been 'packed together perhaps since the creation of the world',[66] or the 'extraordinary shift of ice floes in the seas of the North'[67] cited by people in the Vendée, Deux-Sèvres, Yonne and Landes departments.

The Forests and Climates of the Globe

This way of anchoring climate changes in a global framework received a particularly sophisticated expression elsewhere in Europe at the same time. In the spring of 1823, the Brussels Académie des Sciences et Belles-lettres launched a prize competition to reward the best study of the effects of deforestation on air temperature, winds, and the 'abundance' and 'locality of rainfall'.[68] The initiative was inspired by the French inquiry and obeyed the same kind of concern: assessing the effects of the sale of whole

64 Letter of 25 May 1822 from the Prefect of Eure-et-Loir to the interior minister. 'Eure-et-Loir' file, Académie des sciences collection.

65 Answer table from the Prefecture of Charente (1822). 'Charente' file, Météo France collection.

66 Letter of 18 November 1821 from the advisor to the Gers Prefecture to the interior minister. 'Gers' file, Académie des sciences collection.

67 Report by the Prefect of the Vendée, 18 June 1822. 'Vendée' file, Météo France collection.

68 'Protocole des séances de l'Académie royale des sciences et belles-lettres de Bruxelles', handwritten record, session of 5 May 1823, archives of the Académie, Brussels.

swathes of Belgian national forests by the heavily indebted Kingdom of the Netherlands.[69]

The winner was a Frenchman, Alexandre Moreau de Jonnès.[70] A former officer in the Caribbean, under the Restoration Moreau worked in the cabinet of the Navy minister, in charge of research on agriculture, trade and hygiene in the colonies.[71] His paper opened with a statement: the 'torrent of revolutions' had completed the destruction of the old European forests, already weakened by centuries of population growth and industrial expansion. The balance-sheet was especially alarming in France, and Moreau cautioned against the complete destruction of its woodlands within a half-century.

In order to examine the climatic effects of deforestation, he employed an analytical framework impressive for its innovative character and rigour. The inspiration was a piece by Alexander von Humboldt published in 1817, in which Humboldt invented a new instrument – the isothermal line – tracing curves linking points of the globe with the same average temperature.[72] This seemingly simple innovation would mark the whole future development of atmospheric studies.

But Moreau was above all inspired by the mode of reasoning for which isotherms were the means. For Humboldt, understanding terrestrial climates involved their intellectual 'reconstruction', starting from a basic

69 Ibid., session of 4 May 1823. On these sales, see Pierre-Alain Tallier, 'Ces forêts domaniales qui firent la Société générale (1822–1864)', *Revue belge de philologie et d'histoire*, vol. 80, no. 4, 2002, pp. 1243–74.

70 'Protocole des séances de l'Académie royale des sciences et belles-lettres de Bruxelles', handwritten record, sessions of 6 and 21 May 1825, archives of the Académie, Brussels. Alexandre Moreau de Jonnès, 'Premier mémoire en réponse à cette question: quels sont les changements que peut occasionner le déboisement de forêts', *Mémoires de l'Académie de Bruxelles*, vol. 5, 1825, pp. 1–207. A pharmacist from Mantes, Auguste-Antoine Bosson, also obtained an award with a paraphrase of Rauch.

71 Alexandre Moreau de Jonnès, *Notice sur les travaux scientifiques d'Alexandre Moreau de Jonnès*, Paris: Bourgogne et Martinet, 1842. On the climate of the West Indies, he published *Tableau du climat des Antilles et des phénomènes de son influence sur les plantes, les animaux, et l'espèce humaine*, Paris: Migneret, 1817.

72 Alexander Humboldt, 'Des lignes isothermes et de la distribution de la chaleur sur le globe', *Mémoires de physique et de chimie de la société d'Arcueil*, vol. 3, 1817, pp. 462–602. The map is in Alexander Humboldt, 'Sur les lignes isothermes', *Annales de chimie et physique*, 2nd series, vol. 5, 1817, pp. 102–13 (esp. pp. 112–13). On climatological cartography, see the recent synthesis (and pertinent bibliography) in Sebastian Grevsmühl, 'Visualising Climate and Climate Change: A *Longue Durée* Perspective', in Giuseppe Feola, Hilary Geoghegan and Alex Arnall, eds, *Climate and Culture: Multidisciplinary Perspectives on a Warming World*, Cambridge: Cambridge University Press, 2019, pp. 46–67.

configuration where the Earth is a perfect sphere. In a second phase, he used the isotherms to assess what real climates owed to the influence of 'modifiers': relief, water masses, the superficial condition of the ground, and so on. The idea was an old one, but Humboldt applied it mobilizing a plethora of observations to specify the hierarchy and scope of these 'modifying' influences.

Moreau adopted this approach to analyse the effects of deforestation on the temperature, via its impact on the 'modifiers' (apparent relief, for example, is reduced in the absence of trees). He also sought to map climatic profiles for different regions of the world, so as to evaluate the active impact of trees in each situation. Finally, he borrowed from Humboldt his insistence on quantification. He himself relied on an enormous quantity of data. For his chapter on temperature, he used 30,000 meteorological data, half of which, he wrote, 'belonged' to him.[73]

He ultimately concluded that a forestry and climatic optimum existed for each region of the globe. Below it, one would be approaching the planet's 'end state' – 'vast, sterile, desiccated, solitary spaces'; above it, its 'primitive state' – humid, marshy, and insalubrious.[74] France and Belgium, he warned, were at a tipping point: deforestation must be halted at all costs. The Brussels Academy would warmly compliment this conclusion in its report, which closed with a rebuke to the Kingdom of the Netherlands: think of the responsibility you are assuming in squandering a country's forestry heritage . . .

Forgetting the Inquiry

At the same time in France, Yvart, Huzard and Bosc set about summarizing the ministerial inquiry. They were disconcerted by the responses from the departments. The sometimes succinct, sometimes rambling, character of the submissions; their lack of convergence on the key question of an average alteration of the climate; the near absence of quantitative data – all this contributed to making a diagnosis risky, not to say impossible. The three men delivered their report in February 1824.[75] They deplored

73 Moreau de Jonnès, 'Premier mémoire', p. 30.

74 Ibid., p. 182.

75 *Procès-verbaux des séances de l'Académie des sciences de Paris*, vol. 8, 1824–27, pp. 26–8.

an influx of 'vague theories', terse opinions, and fragile observations. To advance further, it would be necessary 'to request endless explanations that would perhaps require further explanations', and so on ad infinitum. Therefore, they explained, the situation was not propitious to scientific certainty. The only thing to clearly emerge from the inquiry was the fact of the deforestation of mountains – in short, something that was already seen as a fact prior to the circular. Their conclusion, adopted by the Académie as its official response to the government, was that 'it has not found sufficiently positive or complete evidence of contested facts for it to be able to express an opinion.'

An inquiry, then, that fizzled out. An inquiry which, as soon as it was finished, went into the archives where it lay forgotten for nearly 200 years. An inquiry, finally, whose sponsor – the Interior Ministry – did little or nothing. This was because between its launch and its completion, power had changed hands. Decazes and his protégé Mirbel had played a key role in the formation of the Agricultural Council and the production of climate expertise. The replacement of Decazes by Count Siméon in February 1820 did not alter the situation. First, because Mirbel retained his contacts in the Council, which he had handpicked. Then, because Siméon himself took a keen interest in climate questions. His ministerial documents attest to this: they include repeated assents, signed by his hand, to subsidies to Rauch for the diffusion of *Régénération* and the publication of the *Annales Européennes*.[76] It was another matter with his successor, the ultra-royalist Jacques-Joseph Corbière, who deliberately condemned the Agricultural Council to a slow death by no longer calling meetings. The 1821 inquiry likewise fell into oblivion: like an opinion given to a political authority that no longer had any use for it, since it did not have to justify measures it had always opposed. The inquiry was dead and buried.

In another sense, it was no sooner born than it was obsolete. A few months after its conclusion, Moreau de Jonnès triumphed in Brussels with his paper. This scouted out new routes for the study of climate change: an analytical approach to the globe's main dynamics, with an emphasis on numerical data; a distrust of lofty, providentialist systems, specific testimonies, and piecemeal historical evidence. A few years later, when the General Statistics of France was created to equip the country with a

76 National archives, F0/4462.

long-term statistical instrument, Moreau was appointed its director.[77] One of his objectives was to produce a meteorology of France. How? By *silencing* all voices from the provinces with their own opinions on the 'system of their department', and disciplining observers to be mere producers of numbers which the analytical tools at the centre would endow with meaning.[78] In sum, the very opposite of what Rauch was proclaiming when he refused the magisterium of experts and abstract knowledge about the climate, and dreamt of observatories dotted throughout France to produce a climate knowledge rooted in the life of its inhabitants. In the first half of the nineteenth century, the demand for a science by and for the people was very strong.[79] Like the disciples of Charles Fourier or of the doctor François-Vincent Raspail, Rauch yearned for a more democratic form of knowledge of the world, which would also mean conversion to a different relationship to nature. Conversely, Moreau wanted an army of disciplined observers, simple suppliers of numbers. The cacophony, he intimated, must cease. But this noise was also the other side of the capacity, hitherto assumed and partially recognized, to stand as an expert on the climate of one's own environment, its ups and downs, its history.

77 Alp Yücel Kaya, 'Le bureau de la statistique générale de France et l'institutionalisation des statistiques agricoles: l'Enquête agricole de 1836', *Oeconomica*, vol. 3, no. 3, 2013, pp. 421–57.

78 Valentin Pelosse, 'Observation météorologique et sociétés savantes de province, ou la désignation du bon objet scientifique (1821–1878)', Études rurales, nos 118–119, 1990, pp. 69–82.

79 Bernadette Bensaude-Vincent, *L'Opinion publique et la Science*, Paris: Sanofi-Synthélabo/Les Empêcheurs de penser en rond, 2000.

12

The Power of Forests

The redefinition of property forms was a major issue in the opening decades of the nineteenth century. At the heart of the process were forests. Their importance for trade, industry, war, and everyday life put them centre-stage in the great social and political battles over property.

The fight against the commons, ferocious in those years, was waged at least in part in the name of protecting trees and climates. The idea was to bring the rural masses to heel, disciplining the ways they used nature. Woodland commons represented 'a different way of owning' that had to be eradicated.[1] A ferocious struggle also pitted supporters of an absolutist view of individual property against those who wished the state to be able to place checks on what individuals did with their woods.

In play, at the same time, was the resurgence of a way of governing nature that was essential in the history of the state. The forest administration had been essential to the expansion of royal power during the ancien régime:[2] the central authority asserted itself by prohibiting, prescribing, monitoring, and punishing what went on in forests, what was done with trees, dead wood, game, humus and plants. The power of foresters was once again central in the nineteenth century, in the expansion of the nation-state

1 Paolo Grossi, *Un altro modo di possedere: l'emersione di forme alternative di proprietà all coscienza giuridica postunitaria*, Milan: Giuffrè, 1977.

2 Corvol, *L'Homme aux bois*. For an example linking assertion of royal power, strategic issues and forests, see Daniel Dessert, *La Royale, vaisseaux et marins du Roi-Soleil*, Paris: Fayard, 1996, chapter 8.

and the strengthening of its territorial control. Climate change lay at the intersection of these transformations of the state, property, and ways of living with forests. But the power of trees over the climate was uncertain and contested. What was science able to say about it? How were decisions to be made in a situation of doubt and uncertainty? Which should be preferred – laissez-faire or caution?

An Affront to Property

The Revolution had proclaimed the property-owner's absolute power over his forests. A law of 29 September 1791 decreed that private forests would no longer be subject to any form of state control. Thirteen years prior to its formulation by the Civil Code, this new, radical conception of property as individual prerogative, exclusive and absolute, was thus affirmed as regards forestry ownership. But the measure was controversial, and the Napoleonic consular government, in its attempts to reassert the power of the state over social life, sought to curtail it.[3] A law of 9 Floréal Year XI (29 April 1803) amended the law of 1791, beginning with protecting the very existence of forests. Henceforth, any property-owner wishing to clear a parcel of his land had to apply for permission, six months in advance, from the forestry authorities.[4] The latter could oppose it and put the matter to the Finance Ministry, which had the final say.[5] The penalty for any violation was heavy: the owner had to replant at his own expense, and pay a fine.

The law on the authorization of clearance was passed in 1803 for a period of twenty-five years. It was rediscussed in the 1820s, during the debates on the creation of a forestry code. This code, promulgated in 1827,

3 This decision formed part of a wider process of reassertion of the state in matters of forests: the law of 16 Nivôse Year IX reconstituting an autonomous body of forestry agents; the decree of 19 Ventôse Year X subjecting the forests of communes to the 'forestry regime' – that is, to strict regulation. Georges-André Morin, 'La continuité de la gestion des forêts françaises de l'Ancien Régime à nos jours, ou comment l'État a-t-il pris en compte le long terme?', *Revue française d'administration publique*, no. 134, 2010/2, pp. 233–48; Corvol, *L'Homme aux bois. . .*, pp. 239–43.

4 'Loi relative au régime des bois appartenant aux particuliers, aux communes, ou à des établissements publics (du 9 floréal an VI)', *Mémorial forestier*, year XI, pp. 162–7.

5 Exempt were small woods (less than two hectares, and provided they were not in the mountains) and those annexed to a dwelling.

was decidedly unfavourable to communities and their commons.[6] It decreed the abolition of any collective rights over national forests that had not been proven in law (which was very difficult). It allowed the authorities and private owners to abolish traditional user rights by ring-fencing (*cantonnement*) – that is, by conceding a fraction of the woods to the locals and banning them from all the rest. It restricted pastureland. The blow was so hard that it provoked one of the most famous rural revolts of the nineteenth century – the war of the Demoiselles, pitting the inhabitants of Ariège against representatives of the state from the late 1820s.[7] On the other hand, the code greatly benefited private property-owners: as well as the weapon of ring-fencing, they had the right to choose their keepers, and any offences committed in their woods were punished in the same way as in state forests.[8]

There were two main limits to this proprietary absolutism. Firstly, the right of the Navy to take what was needed for their ships. Secondly, the regulation of clearances, renewed after stormy debates. This became a focal point of the struggles waged in the first half of the nineteenth century over private property and the legitimacy of state-imposed constraints on it.

Article 219 of the forestry code confirmed the measure for a further twenty years.[9] During debates on the code, it emerged that from 1821 to 1826, between 2,500 and 8,000 applications for permission to clear had been referred to the Finance Ministry every year.[10] With the rejection rate oscillating between 44 and 87 per cent, between 1,000 and 6,000 operations per year had been prohibited – in other words, a considerable loss for owners wishing to convert their forests to crop cultivation or make a quick profit by liquidating them.[11] Over and above this concrete impediment, for the petty bourgeoisie that had invested in woods, as for

6 Decoq, Kalaora and Vlassopoulos, *La Fôret salvatrice*, Introduction. On the debates surrounding the adoption of the forestry code, see Matteson, *Forests in Revolutionary France*, pp. 187–206.

7 Peter Sahlins, *Forest Rites: The War of the Demoiselles in Nineteenth-Century France*, Cambridge, MA: Harvard University Press, 1994.

8 Matteson, *Forests in Revolutionary France*, p. 197. Andrée Corvol stresses both the prerogatives allowed to the private owners and the reinforcement of the forestry authorities involved in the code: *L'Homme aux bois*, pp. 259–60, 293.

9 *Code forestier*, Paris: Mme Ve J. Dècle, 1827, pp. 72–3.

10 Jacques Curasson, *Le Code forestier conféré et mis en rapport avec la législation qui régit les différents propriétaires et usagers dans les bois*, vol. 2, Paris: Gauthier frères et Cie, 1828, pp. 193–4.

11 Unfortunately, no more precise data has survived concerning this refusal.

the major forestry interests, the need for authorization was an intolerable infringement of the idea of outright ownership.

Forestry Externalities

For French liberals, article 219 became the emblem of the vexations inflicted on property-owners by the state. It was wickedly unfair, they pointed out: the state profited from the ban on land clearance by selling parcels from the national forests at a higher price, accompanied by authorization. The issue was a topical one, for the beginnings of the July Monarchy saw a second wave of sales, following those of the Restoration.[12]

Article 219 led to a war of attrition in the Assembly. The crusade for private property was led by Alexandre Anisson-Duperron, former director of the National Printing-House, a major landowner and promoter of liberal economic doctrines.[13] He went onto the offensive in 1834, tabling an amendment stipulating that land clearance could only be forbidden in forests located in mountains or on sand dunes.[14] This implied that clearances would be freely allowed elsewhere – in other words, it was a deregulation involving all the great lowland forests, by far the best in terms of land values and prospects for cultivation. The liberals and major property-owners (who were often the same people) attacked at an angle, lauding the virtues of article 219 and the ecological benefits of forests – but only in the case of mountainous areas, so as to win the battle on the plains.

The liberal position was justified in the first instance by a hierarchy of properties. Charles Comte, the son-in-law of Jean-Baptiste Say and a leading figure of French liberalism, expounded this argument in his influential *Traité de la propriété*.[15] Because lowland properties, Comte explained, were 'infinitely more valuable' than highland ones, 'any action taken by government to preserve springs and rivers' (by prohibiting the clearing of slopes) was legitimate, since it was conducive to ensuring

12 116,780 hectares whose sale was organized by a law of 25 March 1831.

13 Alexandre Anisson-Duperron was the author of a free trade handbook, *De l'affranchissement du commerce et de l'industrie*, Paris: Librairie universelle de P. Mongie, 1829. In the 1840s he participated in Frédéric Bastiat and Michel Chevalier's Association pour la liberté des échanges.

14 *Journal des débats*, 20 February 1834 and 4, 6 and 25 May 1834.

15 Charles Comte, *Traité de la propriété*, 2 vols, Paris: Chamerot/Ducollet, 1834.

the essentials of economic wealth.[16] The argument also had a historical corollary. Because the land was more fertile on the plains, that was where properties had first been created. This anteriority implied protecting them first and foremost.

But why not leave that to market mechanisms? To explain it, Comte advanced a very modern idea: the 'eco-systemic service' performed by the mass of trees. The term is not present, but the idea is. He began with a critique of Arthur Young, who, in *Travels in France*, maintained that a government must do nothing to prevent land clearance when the price of wood was too low for forestry properties to be valued at the same level as crops.[17] In effect, according to Comte, forest matters were very specific and invalidated the usual reasoning about the pursuit of interest and market mechanisms. He declared that mountain forests rendered services to the lowlands that the market *did not take into account* in monetary form – externalities. These services consisted in the activity carried out by mountain trees *remotely* in supplying underground water tables, protecting the plains from floods and droughts, and regularizing flows, thus enabling irrigation.

However, nothing in market logic motivated individual actors to 'perpetuate the life-span' of these trees, for the services they rendered were neither sold nor bought (and indeed, frequently went unperceived). 'Property-owners', wrote Comte,

> *cannot be paid* for the services rendered by their forests to the populations living in the basins of rivers, through the influence they have on the distribution of water. They can expect no profits other than from the sale of wood, and it is only natural that they constantly compare the income they earn from that with what the same land would yield were it to be cleared. . . . *Forests or woods, especially in certain locations, thus render services to the nation which offer no particular advantage to their owners – services that everyone enjoys, while no one has the will or the power to pay for them in order to perpetuate their life-span* (our emphasis).[18]

16 Ibid., vol. 1, p. 220.

17 According to him, such measures arbitrarily favoured the interests of consumers at the expense of those of property-owners.

18 Comte, *Traité de la propriété*, pp. 233–4.

According to Comte, this was an exceptional case, in which the state must intervene. He leaned towards tax incentives for property-owners to encourage conservation.[19] Other thinkers, such as Charles Dunoyer, envisaged regulatory measures reserved for mountain woods.[20] The most radical view was that of the ex-Saint-Simonian and penman of French liberalism, Michel Chevalier: to regularize the flows of streams and rivers – and hence facilitate the general flow of trade – he thought it indispensable to reforest mountainous areas.[21] And he went further: 'Deforestation is a conquest of nature by Man; woods must disappear from the plains and give way to crops.' The country must thus be divided in two: in the Alps, the Vosges and the Pyrenees, mountain forests continue to regulate flows and supply timber; in the plains – wherever possible – there should be agriculture. This also involved a conception of energy: for Chevalier, France's industrial trajectory must combine cheap wood from the mountains and an iron industry, whereas England used coal because it no longer had any forests.[22]

Faced with these proposals for tree protection in the mountains but not in the plains, supporters of regulation warned against the wave of clearances that would ensue and the impact on hydrological cycles. They predicted that the low financial return on forestry capital, which was the case under the July Monarchy,[23] would drive owners to chop down their woods: the public authorities must take a stand against this impending disaster.

Playing on Uncertainty

In 1834, the liberals failed to get their liberalizing measure passed in the Assembly. But they returned to the attack the following year.[24] The

19 Ibid., pp. 249–50.

20 Charles Dunoyer, *De la liberté du travail*, vol. 2, Paris: Guillaumin et Cie, 1845, pp. 452–3.

21 Michel Chevalier, *Des intérêts matériels de la France*, Paris: Gosselin et Coquebert, 1838, pp. 191–3. See also his *Cours d'économie politique fait au Collège de France. Deuxième année (1842–1843)*, Paris: Capelle, 1844, pp. 422–5.

22 'France. Travaux des ingénieurs des mines (compte rendu de 1837)', *La France industrielle, manufacturière, agricole et commerciale*, 16 September 1838.

23 This low profitability was bound up in particular with increasing competition from coal and transport problems. Raymond Lefebvre et al., *Les Eaux et Forêts du XIIe au XXe siècle*, Paris: Éditions du CNRS, 1987, p. 493.

24 *Journal des débats*, 22, 23 and 24 January 1835. Anisson-Duperron this time proposed a reform that confined article 219 to a closed list of situations: those where the

debates were vicious. The reform was contested by the Baron de Ladoucette, former prefect under the Empire and a specialist in agricultural questions.[25] Why facilitate forest clearance, he demanded, when there was a vast amount of idle land in France that could be cultivated? And was it right to disregard the defensive function of woods at frontiers, where they served as natural bulwarks against the enemy? Not to mention the role played by forests against erosion, of course, which Ladoucette, a former prefect of the Hautes-Alpes, knew all about.

After several days of debates, each article of the proposal had been adopted, but then a dramatic volte-face occurred: the vote on the text as a whole was negative. The pro-regulation camp breathed a sigh of relief, but could sense which way the wind was blowing. In 1836 it would regroup around a new line of defence. So liberals sought to focus attention on the ecological services performed by mountain forests? Easy! Facing them, Ladoucette and Passy, the minister of Public Works, demanded universal protection since, they contended, *all* forests, *in the plains* and mountains alike, maintained *the climate of the whole country* at a point of equilibrium. In a vibrant speech to the Chamber, Passy invoked Egypt and Puerto Rico, where plantations and the protection of mountainous areas were said to have restored the rains.[26] Scientists had to be mobilized; he therefore proposed adjourning the debate long enough for a working group to come up with a diagnosis. There was uproar: 'An inquiry is a burial certificate!' remonstrated a deputy.

At this point François Arago mounted the rostrum.[27] This speaker was not only director of the Paris Observatory and a world-famous astronomer, he was also the patron of French science – a media personality whose words were carried by the press and peddler literature deep into the country. And he was a politician, a deputy close to the centre-left and the Thiers government. In his view, the state must protect forests.

He stressed the close links between forest and climate. We cannot, he said, predict what will happen if French forests decline further. Before the great deforestation that accompanied the country's agricultural

administration could *prove* that problems of soil erosion, conservation of springs, defence against winds, sands, downpours or avalanches arose.

25 *Notice biographique sur le Baron de Ladoucette. Société nationale et centrale d'agriculture*, Paris: Imprimerie de Mme Vve Bouchard-Huzard, n.d.

26 *Journal de débats*, 28 February 1836. The debates had been revived by a fresh proposal for reforming article 219.

27 Ibid.

development, the climate in France was 'excessive' – the seasons were more extreme. Deforestation had therefore been beneficial. But could one be sure that the balance was not being tipped too far, at the risk of precipitating disaster? Arago stressed the reigning uncertainty that prevented a rational decision. He insisted that 'precise, incontestable ideas' were needed for the Assembly to make up its mind.[28] The scientific commission proposed by the minister must be set up. The astronomer's voice prevailed: parliamentary discussion was adjourned while the experts were consulted.

A commission was created with seventeen members, including Anisson-Duperron, Ladoucette, Arago, the physicist Gay-Lussac, and parliamentarians from both sides.[29] Meeting until the summer of 1836, it finally proposed to submit to prefects and the Académie des sciences a new survey on climate change, 'from historic times onwards'. The questionnaire focused on the evolution of temperatures, rains, snows, winds, river spates, and springs.[30] As in 1821, the Académie des sciences was tasked with a summary. For his part, Arago tried to mobilize Orientalist experts, philologists and historians, approaching the Académie des inscriptions et belles-lettres to request its help researching 'the former thermometric or climatological condition' of Europe in Greek, Roman and Oriental authors.[31]

Eighteen months later, the question of clearances was back under discussion in the Chamber.[32] Here was an opportunity for Arago to reproach the government. According to him, the survey had never been forwarded to its addressees: 'The questions remained in boxes.'[33] In the interim, the fall of the Thiers cabinet at the start of September 1836 had altered the situation and cut Arago off from his contacts at the top of the state. Nor was the astronomer able to get hold of the responses to the inquiry of April 1821. He implicitly accused the government as he ironized: 'I did not think I would have to go and look for them at the grocer's, but perhaps I would have found them.'[34]

28 Ibid.

29 Order of the King, 29 March 1836, *Bulletin des lois*, 9th series, vol. 12, 1836, pp. 104–5.

30 'Physique du globe. Questions relatives aux effets des défrichements', *Comptes rendus de l'Académie des sciences de Paris*, vol. 3, 1836, pp. 398–9 (session of 26 September 1836).

31 'Physique du globe. Influence des défrichements', ibid., p. 768 (session of 19 December 1836).

32 *La Presse*, 4 March 1838.

33 Ibid., 6 March 1838.

34 Ibid.

On the substance, Arago remained categorical: 'The matter has been debated by the most skilled meteorologists: nearly all of them believe in the considerable climatological influence of woods located in the plains.'[35] As to its exact mechanisms and effects, Arago confessed: 'I was, and remain, in doubt'. But what mattered was the immensity of the risk – nothing less than the destruction of France's climate and agriculture. Prudence must prevail and uncertainty prompt protection. One cannot dice with the fate of the Nation.

Arago advocated a fresh adjournment of the debates until the inquiry was complete. The deputy Jaubert was outraged by this: another 'meteorological adjournment',[36] when science (and Arago) had proved unable to say anything more on the subject in eighteen months? Was there not quite simply nothing to be known, nothing to be discovered? But after the physicist Gay-Lussac had intervened to contradict Arago, by relativizing the influence of lowland forests ('That is a mistake',[37] retorted Arago), the adjournment was put to the vote and adopted. The astronomer's strategy, his appeals to caution in the face of uncertainty, had won the day. Forests were not a piece of property, to be disposed of at whim. As for the inquiry, it was never completed. In 1844, the Finance Ministry chased up the Académie des sciences, which was due to summarize the results. It dodged the issue, saying only that 'highly delicate' matters were involved.[38]

Return to Tacarigua

Arago's position on climate change had changed considerably since the 1820s. Back then he still tutted about the 'public' which, no sooner did the thermometer exceed its usual limits, was convinced that 'it had never been observed so high or so low'.[39] Now, he accepted the possibility of large-scale anthropogenic change. In 1834, in an article that was widely

35 François Arago, Œ*uvres complètes. Mélanges*, Paris: Gide, 1859, p. 442.

36 *Journal de débats*, 6 March 1838.

37 Ibid. Maurice Crosland mentions the influence that Gay-Lussac's status as a landowner might have had on his position: *Gay-Lussac: Scientist and Bourgeois*, Cambridge: Cambridge University Press, 1978, pp. 246–7.

38 'Correspondance', *Comptes rendus de l'Académie des sciences de Paris*, vol. 19, 1844, p. 403 (session of 19 August 1844).

39 François Arago, 'Tables des températures extrêmes observées à Paris et dans d'autres lieux du globe', *Annuaire du Bureau des longitudes* for 1825, pp. 164–78 (p. 164). He then supplied a table of extremes of temperature in Paris from 1665 to 1823, as a

distributed, including abroad, he discussed the issue at length, relying on two sources: temperature measurements, dating back to the seventeenth century, by Florence's Accademia del Cimento; and land registry archives, plus the dates of remittance of ground rent (in wine), for the Vivarais region.[40] These data seemed to indicate that summers had become less hot and winters less cold.[41] Were this to be confirmed, no more doubt, it could only be explained by human action – and this despite the fact that men's works 'barely penetrate the epidermis of our globe'.[42]

In those years Arago occupied an institutional and mediatic position unique in the history of science in France. He directed the Paris Observatory and dominated the Académie; his interventions filled the newspapers; he controlled the *Annales de chimie et de physique*, one of the country's principal scientific journals.[43] Thus he heavily influenced both the scientific consensus and the public opinion of his time.[44] In the midst of the debate over climate and forestry policy, he would use this position of power to get one of the key texts on thinking about climate change in the nineteenth century published.

At the beginning of the 1820s, Simón Bolívar commissioned the naturalist Francisco Zea to recruit personnel in France for the brand-new School of Mines in the state of Gran Colombia. A young graduate from Saint-Étienne's École des Mines, Jean-Baptiste Boussingault, was selected and embarked for South America. Boussingault also had scientific ambitions and a powerful patron – Alexander von Humboldt – who helped him prepare a research programme, gave him instruments, and introduced him to Parisian scientific society.[45] In 1823, Boussingault stayed on the shores

rational benchmark on which to base an opinion – a sound one, being scientific and quantitative – on the seasons.

40 François Arago, 'Sur l'état thermométrique du globe terrestre', *Annuaire du Bureau des longitudes* for 1834, pp. 171–240.

41 In the Vivarais, wine harvests used to be earlier and the vineyards went further up the hillsides.

42 Arago, 'Sur l'état thermométrique du globe terrestre'.

43 Bruno Belhoste, 'Arago, les journalistes et l'Académie des sciences dans les années 1830', in Patrick Harismendy, ed., *La France des années 1830 et l'Esprit de* réforme, Rennes: Presses Universitaires de Rennes, coll. 'Carnot', 2006, pp. 253–66.

44 On Arago as a public figure, see Theresa Levitt, '"*I thought this might be of interest* . . .": The Observatory as Public Enterprise', in David Aubin, Charlotte Bigg and Otto Sibum, eds, *The Heavens on Earth: Observatories and Astronomy in Nineteenth-Century Science and Culture*, Durham, NC: Duke University Press, 2010, pp. 285–304.

45 Frederick McCosh, *Boussingault, Chemist and Agriculturalist*, Dordrecht/Boston/Lancaster: D. Reidel, 1984, pp. 24–5.

of Lake Tacarigua in Venezuela.[46] There was nothing fortuitous about this: the lake had been familiar to scientists since it had been studied by Humboldt in the account of his famous expedition to America.[47]

In 1800, some locals told him that its level had dropped over time and Humboldt conducted an investigation. His conclusion? It was human action in the form of land clearance that had dried up the tributaries of Tacarigua and caused its decline. Three years later he made the same diagnosis in Mexico City: 'The city built in the middle of a lake' that had dazzled the first Spanish conquerors was no longer to be found.[48] As in Venezuela, the fault was the colonizers' frenzy for profit and destruction: in Humboldt, the science of nature fuelled a critique of colonization. In his travel diary, he wrote: 'Far away from home, Europeans are as barbarous as Turks, even more so because they are more fanatical.'[49]

His interpretation of the desiccation of Tacarigua was not so much climatic as pedological. For him, trees acted by regulating evaporation from the soil, stabilizing the humus that served as a buffer zone for run-off. He borrowed this schema from an engineer from Ponts et chaussées, Gaspard de Prony, who in his book on the hydrography of the Pontine Marshes deployed it to oppose arguments in terms of 'affinities between vapours and forests', or 'other more or less hypothetical principles'.[50]

Returning to Tacarigua in spring 1823, Boussingault followed in Humboldt's footsteps at his request. Twenty years on, he confirmed the link his predecessor had made between deforestation and desiccation. The

46 On Boussingault, see ibid.; Jean Boulaine, 'Boussingault', in *Dictionnaire des professeurs du Conservatoire des arts et métiers*, vol. 1, CNAM, 1994, pp. 246–63; Gregory Cushman, *Guano and the Opening of the Pacific World*, Cambridge: Cambridge University Press, 2013, pp. 33–40; file on 'Jean-Baptiste Boussingault' in the archives of the Académie des sciences de Paris.

47 *Voyage de Humboldt et Bonpland. Première partie. Relation historique*, vol. 2, Paris: N. Maze, 1819, pp. 67–77, and Alexander Humboldt, *Voyage aux régions* équinox*iales*, vol. 5, Paris: N. Maze, 1820, pp. 162–86. On Humboldt and the links between trees, climate and water cycle, see Engelhard Weigl, 'Wald und Klima: Ein Mythos aus dem 19. Jahrhundert', *HiN. Internationale Zeitschrift für Humboldt-Studien*, vol. V, no. 9, pp. 80–99; Gregory T. Cushman, 'Humboldtian Science, Creole Meteorology, and the Discovery of Human-Caused Climate Change in South America', *Osiris*, vol. 26, no. 1, 2011, pp. 16–44; Frank Holl, 'Alexander von Humboldt und der Klimanwandel – Mythen und Fakten', *HiN. Internationale Zeitschrift für Humboldt-Studien*, vol. XIX, no. 37, 2019, pp. 37–56.

48 Alexander Humboldt, *Essai politique sur le royaume de Nouvelle-Espagne*, vol. 1, Paris: F. Schoell, 1811, Book 11, chapter 8.

49 Quoted in Cushman, 'Humboldtian Science', p. 30.

50 Gaspard de Prony, *Des marais pontins*, Paris: Imprimerie nationale, 1818, p. 325.

issue was a sensitive one and, for a mining engineer like Boussingault, anything but abstract. As the manager of a gold mine in Marmato, Colombia from 1826–30, Boussingault was confronted with the choice between logging – necessary for wood charcoal and facilities – and the availability of running water to operate the machines.[51] He had his subordinates perform meteorological observations, for the facts were clear: with the mine in operation, the rivers had less water flow and that meant less gold.

This whole corpus of observations would be handed over to Arago by Boussingault, after his return to France, to fuel his forestry crusade. Arago, whom he had met while preparing his journey to America, had urged Boussingault to publish him in his journal *Annales de chimie et de physique*. Fifteen years after Tacarigua, Boussingault dug up his notes and wrote an article on the link between rainfall and forests which, translated into English, attracted considerable interest.[52] From the observations around the Marmato mine, he concluded that local land clearances impacted on the soil, causing running water to evaporate, which could have serious consequences for extractive industries.[53] Above all, on a different level, he affirmed, it might cause climate change: 'Major clearances *reduce the annual quantity of rain* that falls on a country.'[54] As evidence of this he made a comparison between the large equatorial forests and the arid deserts that occupy the west coasts of Colombia and Ecuador: their geographical positions being analogous, only the presence (or absence) of forests gave them such utterly dissimilar climates.

Boussingault's words went around the world, especially as he integrated his conclusions on climate change into his treatise on rural economies, which would soon be part of the canon of global agronomy in its French, German, English and Italian versions.[55] Boussingault's theses struck a particular chord because they enjoyed peerless scientific backing: that

51 On his experience in Marmato, cf. Jean-Baptiste Boussingault, *Mémoires*, vol. 4 (1824–30), Paris: Chamerot et Renouard, 1903.

52 Jean-Baptiste Boussingault, 'Mémoire sur l'influence des défrichements dans la diminution des cours d'eau', *Annales de chimie et de physique*, vol. 64, 1837, pp. 113–41; 'Memoir Concerning the Effect which the Clearing of Land has in Diminishing the Quantity of Water in the Streams of a District', *The Edinburgh New Philosophical Journal*, no. 24, 1838, pp. 85–106.

53 Ibid., pp. 136–8.

54 Ibid., p. 141.

55 Jean-Baptiste Boussingault, Économie rurale: considérée dans ses rapports avec la chimie, la physique et la météorologie, Paris: Bréchet Jeune, vol. 2, 1844, pp. 701–36 (translated into English in 1845, Italian in 1846 and German in 1851).

of von Humboldt who, at the start of the 1840s, declared that he now believed in the large-scale climatic impact of trees.[56] English foresters in India, George Perkins Marsh in the United States, and various geographers and climatologists until the end of the century – all found in them the diagnosis they sought on anthropogenic climate impact.[57]

Political struggles over forestry property continued to rage, as article 219 survived attacks on it until the Second Empire.[58] Arago played a key role in providing a bulwark against liberalization, with the weapon of expertise and the argument of uncertainty. But this success also stemmed more broadly from the rise to power in the second quarter of the century of a powerful, organized 'forestry party', militating for state control over forests.

From the 1830s, the forestry authorities began to recover their pride, after the sombre decades following the Revolution. They were increasingly busy on the ground and influential within the state apparatus. The forestry code gave them a stable framework of juridical action and the École forestière of Nancy (founded in 1824) fostered the emergence of a strong corporate culture. State forestry now benefited from powerful connections to politicians, civil servants, journalists, and scientists won over to its cause (and often advocates of a technocratic, developmental state).

Belief in anthropogenic change was part and parcel of the professional culture of foresters. It had become entrenched during the 1810s and 20s, under the influence of Rauch, Dugied, and Jacques-Joseph Baudrillart – a theoretician of the forest who was behind both the code and the Nancy school.[59] The theory of anthropogenic climate change became part of the everyday baggage of forestry managers, at a time when a 'technical turn' distanced them from law and they began to train more frequently as

56 Alexander Humboldt, *Asie centrale. Recherches sur les chaînes de montagnes et la climatologie comparée*, vol. 3, Paris: Gide, 1843, pp. 198–206. See also from 1831, but with a damper on the existence of a supra-local impact, Alexander Humboldt, *Fragments de géologie et de climatologie asiatiques*, Paris: Gide, vol. 2, 1831, pp. 550–6.

57 Cf. Chapter 15 in this volume.

58 The provision was renewed in successive debates and votes in Parliament in 1847, 1850, 1851, 1853, 1856: *Journal des débats*, 22 July 1847, 23 July 1850, 23–24 July 1851, 12 May 1853, 23 June 1856.

59 Jacques-Joseph Baudrillart, *Nouveau manuel forestier: à l'usage des agents forestiers de tous grades*, Paris: Arthus-Bertrand, 1808, vol. 2, pp. 403–5; 'Notice historique sur les forêts', *Annales forestières*, vol. 24, April 1810, pp. 168–81; and 'Mémoires sur l'aménagement des forêts', *Annales forestières*, vol. 44, December 1811, pp. 529–68.

engineers instead.[60] From the late 1820s to the 1860s, the pupils who arrived at the Nancy school were introduced in one of their first lectures to the knowledge of trees, stressing their effects on the climate and 'the utmost importance for the climatic condition of a country that the forests there be suitably distributed'.[61] The corollary was the indispensable resources required by the forestry technocracy to create these conditions.

Today it is difficult to imagine the power of the forestry camp at a time when wood was the 'petrol' and 'concrete' of societies.[62] Climate change was a vector of the resurgence of this formidable social and political force, destabilized by the revolutionary period. Having been a defensive argument against liberalization, climate action would soon become an offensive weapon. First when, in the wake of decades of warnings, the French forestry administration mobilized a wealth of resources to control mountain environments. Then, when they made the colonies a land of conquest to the same end: imposing their political plan for a rationalized environment, an ordered rural society, and the control over popular ways of using nature.

60 On this 'technical turn' with the creation of the École forestière de Nancy, cf. Corvol, *L'Homme aux bois*, p. 246.

61 See the classic course by Bernard Lorentz and Louis Parade, the first two directors of the school: *Cours élémentaire de culture des bois*, Paris: Huzard et Nancy, George-Grimblot, 1837, pp. 15–17, and then the 1855 edition, pp. 17–20, and the 1867 edition, pp. 21–4.

62 As Jérôme Buridant put it – that is, a prime source of energy and an indispensable building material.

13

The Horizon Clears

After 1850, France became covered with railways and steam engines, telegraph lines, roads, and a vegetation infrastructure of forests planted to stabilize mountains, dunes and waterways. The country entered technological modernity, a modernity of infrastructure and networks that profoundly altered the economic system, lifestyles, and ways of governing. The state was the driving force in these changes. Its engineers, scientists and foresters refashioned the country in the name of rationality, productivity and security. By the same token, they strengthened the grip of central government over land and society.

Space contracted: people communicated and moved around at high speed, courtesy of the telegraph and train. The transport revolution opened up new, vast prospects for agricultural, forestry and mining businesses. Resources, products and human beings were readily available: trade density – regional, national, global – skyrocketed. Everything was reshuffled, from economic links to space.

Anxiety about climate change became red-hot in the decades of the mid-nineteenth century. To safeguard towns and trade flows, the state and its foresters tackled mountains, with a view to restoring the major water cycles. Over the long term, however, the new forms of technological activity helped push back the spectre of climate deterioration. Contemporary global change starkly reveals the limitations of our industrial development. At that time, by contrast, the growing wave of improvements and networks was instrumental in relegating the idea of human climate action.

Repairing France: From the Sky to the Ground

The 1840s and 50s were marked in France by a succession of high tides and dramatic floods, which made controlling waterways a national priority. These disasters prompted debates and controversies in scientific, technical, political and media circles. But, on one crucial fact, there was a consensus: the human responsibility for these disasters, even if everyone defended their own notion of the mechanisms involved.[1]

In late October and early November 1840, the Rhône and the Saône burst their banks, causing enormous damage in Lyon and Avignon. The Camargue was submerged. In October 1846, it was the turn of the Loire and the Allier to experience huge spates due to storms and flash floods in the Cévennes. Riverbanks were devastated, bridges swept away, and cities (Tours and Roanne) inundated by water. The memory of these catastrophes was still vivid when, in spring 1856, a rare flood event affected the basins of the Rhône, the Loire and the Saône, or two-thirds of the country. The material damage was considerable.[2]

Several decisions were taken to deal with the threat. Warning systems for rising waters were put in place, which harnessed the speed of the telegraph by linking towns to rainfall-measuring stations upstream.[3] In 1856, the government also decided to create specialist study and work services operating in the basins of the Loire, the Rhône, the Garonne and the Seine. Engineers mapped the basins and assessed the condition of structures and their effectiveness. These forms of expertise and intervention were the fruit of a profoundly political reflex, proclaiming the country's simultaneously 'political' and 'natural' stability, and ensuring it in practice.[4] The Empire sought to assert itself by rectifying society and environment

1 For a different reading of these disasters, construed in the light of the emergence of environmental awareness in nineteenth-century France, see Caroline Ford, *National Interests*, chapter 3.

2 Denis Cœur, 'Les inondations de mai–juin 1856 en France: Dommages et conséquences', *La Houille blanche. Revue internationale de l'eau*, no. 2, 2007, pp. 44–51; Jacques Bethemont, '1856: De la gestion d'une catastrophe au bon usage d'une crise', *La Houille blanche. Revue internationale de l'eau*, no. 1, 2007, pp. 22–32; Annie Méjean, 'Utilisation politique d'une catastrophe: Le voyage de Napoléon III en Provence durant la grande crue de 1856', *Revue historique*, no. 597, 1996, pp. 133–52.

3 Or downstream of Lyon, for the Saône: Dr Lortet, 'Notice sur la commission hydrométrique de Lyon', *Annuaire météorologique de la France*, 1849, pp. 362–6.

4 To offer reassurance, Napoléon III also visited the victims, in a glare of publicity. On these aspects, see Méjean, 'Utilisation politique d'une catastrophe'.

Bibliothèque municipale de Lyon/P0546 S 0105

Flooding of Lyon in 1856: view of the quai Saint-Antoine, quai des Célestins and quai Tilsitt. Photograph by Louis Froissard, 18 May 1856.

alike, thanks to science. In a speech in February 1857, Napoleon III said so virtually word for word: everything 'gives me hope that science will succeed in taming nature'. Furthermore: 'I make it a point of honour that in France rivers, *like the revolution*, will return to their bed and not leave it again' (our emphasis).[5]

How were these catastrophes, which twice in ten years had drowned French towns, to be interpreted? The waves of floods in 1846–56 acted like detonators sparking fresh doubt: what if the destruction of forests – and thus of the climate – was the active cause of this large-scale disruption of nature? In November 1846 a doctor, Joseph-Jean-Nicolas Fuster, sounded the alarm with three successive articles in the paper *L'Univers*.[6] 'The sole cause of the flooding is the denudation of our central mountains', he wrote. He proceeded to explain the functions of trees, which broke up clouds, distributed rainwater and averted storms. Any response in the form of building embankments and dikes, he added, would be a waste of time and money if forests were not protected and reforestation implemented. An even, mild climate – France's of yesteryear – must be restored. Fuster's assessment was sombre: the 'labours of several generations will be required'. His warnings were taken up in the provincial press, as in Dijon's

5 *Journal historique et littéraire*, vol. 23, 1856–57, p. 541 ('Discours prononcé par l'empereur des Français').

6 'Des inondations de la Loire', *L'Univers*, 1, 3, and 6 November 1846.

Le Spectateur, which carried the headline: 'Restore France's Forests? A Matter of Life or Death.'[7]

Playing out in the background was a violent struggle for influence in the state apparatus. It opposed two technocracies, vying with each other to take advantage of the crisis. 'Since the floods of 1856, how many have set to work to find remedies for the terrible scourge!', wrote an ironical observer. 'And all insistently demand that the state, with its treasury and its omnipotence, immediately apply *their* infallible means for preventing floods' (our emphasis).[8] On one side stood the powerful Ponts et Chaussées administration, which saw an opportunity to establish its hegemony over the waterways. Its engineers beat loud drums for the solutions of diking and artificialization, in uplands and lowlands alike. At the same time they unanimously denied that deforestation was the prime cause of flooding.[9] One of them quipped: so you want to plant trees all over the place and 'recreate Gaul in the nineteenth century'?[10] In the opposite corner, the forestry technocracy saw these catastrophes as the perfect opportunity to promote its long-standing plans for more regulation on forest use and for reforesting mountains. It encouraged supporters to disseminate the interpretation of rising water levels as the consequence of deforestation at altitude.[11] They could draw on countless reports, articles and pamphlets which had, for several decades, been depicting French mountains as wrecked by deforestation. The argument raged between the Ponts et Chaussées administration and the forests authorities to sway the arbitration of imperial power and obtain ever greater resources.[12]

More worryingly, some challenged the great idea of a progress whose feats the regime had just celebrated in the 1855 Universal Exhibition in

7 *Le Spectateur*, Dijon, 19 November 1846.

8 Raudot, 'Les inondations', *Journal d'agriculture pratique*, October 1856, pp. 269–76.

9 See, for example, Aristide Dumont (engineer from Ponts et Chaussées), 'Des inondations et des moyens de les prévenir', *La Presse*, 21 November 1846; François Vallès (engineer from Ponts et chaussées), *Études sur les inondations, leurs causes et leurs effets*, Paris: V. Valmont, 1857.

10 Dumont, 'Des inondations et des moyens de les prévenir'.

11 See, for example, Jacques Valserres, 'Les inondations', *Annales forestières*, January 1856, pp. 21–5; L. Hun, *Des inondations et des moyens de les prévenir*, Paris: Bureau des Annales forestières, 1856; Auguste de Gasparin, 'La question du reboisement et des barrages', *Journal d'agriculture pratique*, November 1856, pp. 353–6.

12 A.F. d'Héricourt, 'Les inondations et le livre de M. Vallès', *Annales forestières*, December 1857, pp. 310–21.

Paris. Eugène Huzar, who had already created a sensation with his alarmist work, *La Fin du monde par la science*, made the most of the floods to repeat his attack.[13] *L'Arbre de la science* (1857) opened with a pathos-filled description of the flooding of the Loire and the Rhône.[14] These calamities were presented as confirmation of his theses: any technological and industrial progress (deforestation, railways, vaccination) inexorably led to catastrophe. Similarly, in the same year the botanist and horticulturalist Élie-Abel Carrière published a book informed by the writings of Bernardin de Saint-Pierre, Rauch, Rougier de la Bergerie, and the ultra-royalist deputies of the 1810s,[15] hailing them as prophets of the ongoing doom. Carrière described an industry that 'depopulates the countryside', 'enfeebles the mind and body alike', 'devours its children and lives exclusively off their death'; a progress ' in name only', since it destroyed nature and stole it from men, devouring forests and climate.[16] For Huzar and Carrière, the recent disasters should not encourage further extension of the sway of technology over men and things. On the contrary, the deadly race to destruction and servitude had to be stopped.

For its part, the government stuck to a strictly technical and developmentalist analysis. The finger was straightaway pointed at the damage to mountains. In July 1856, in a letter to his minister of Public Works, Napoleon III wrote: 'Where does the sudden flooding of our major rivers come from? From the water that has fallen on the mountains: very little derives from the water fallen on the plains . . . the ground, mostly composed of rocks or gravel, does not absorb the water, while the steepness of the slopes sends all the fallen water into rivers, whose level suddenly rises.'[17] But who was to be entrusted with repairing the mountains, as well as acting in the plains, as many were demanding? The merciless struggle between foresters and engineers begat the overarching triumph of the technological state. Ponts et chaussées annexed the management of basins

13 Eugène Huzar, *La Fin du monde par la science*, Paris: E. Dentu, 1855. Cf. Jean-Baptiste Fressoz, 'Eugène Huzar et l'invention du catastrophisme technologique', *Romantisme*, no. 150, 2010/4, pp. 97–103.

14 Eugène Huzar, *L'Arbre de la science*, Paris: E. Dentu, 1857.

15 Élie-Abel Carrière, *Les Hommes et les choses*, Paris [self-published], 1857.

16 Quotations taken from the title page of Carrière.

17 'L'Empereur adresse la lettre suivante au ministre des Travaux publics' in Plombières, 19 July 1856, in *Le Moniteur universel*, 21 July 1856. The letter was also printed in the local press.

and high tide warnings; the foresters extended their sway to become the embodiment of the technocratic state in the mountains.

Mountains became the theatre of operation of a forestry authority with exorbitant powers.[18] The objective was twofold: conservation of highland forests and reforestation. In the tradition of discourse since the eighteenth century, agro-pastoral populations were identified as the guilty parties in the degradation of forests and eco-systems, which was in turn the cause of flooding in the lowlands.[19] The law of 28 July 1860 provided for support for planting, but also measures of compulsory reforestation imposed on private parties on pain of expropriation.[20] The Third Republic would continue this policy with its major programmes of 'restoration of mountainous terrain', in vigorous but less coercive forms.[21] The outcome was a veritable 'conquest of the mountain' by the forestry authorities, the artificialization of mountain forests, and the technical intensification of logging. As Bernard Kalaora has emphasized, this movement sprang from the desire to modernize and rationalize space, which made sense in the context of country-wide industrialization and network-building.[22] As regards the national estate, 250,000 hectares of woods fell into the hands of the state between 1860 and 1914, thanks to the more or less compulsory sales and expropriations accompanying the conservation and reforestation measures.[23]

From 1860–80, this mountain offensive and the confrontation with Ponts et chaussées elicited from foresters a vast output of expert opinion and adoption of positions on the links between forests, water cycle and flooding. This focused on two related but distinct topics: on the one hand, the impact of trees on the atmosphere and precipitations; on the other, their effect on the draining of water flows into the ground. This distinction was clearly perceived by contemporaries. Alexandre Surell's *Étude sur les*

18 There is an abundant historical literature on the activity of foresters in the mountains. See Guillaume Decoq, Bernard Kalaora and Chloé Vlassopoulos, *La Forêt salvatrice: reboisement, société et catastrophe au prisme de l'histoire*, Seyssel: Champ Vallon, 2016; Tamara Whited, *Forests and Peasant Politics in Modern France*, New Haven: Yale University Press, 2000; Lefebvre et al., *Les Eaux et Forêts du XIIe au XXe siècle*; Corvol, *L'Homme aux bois*.

19 Martine Chalvet, 'Reboiser la forêt provençale. L'instrumentalisation de Surell par les élites provençales', *Annales des Ponts et chaussées*, no. 103, 2002, pp. 44–50.

20 It was complemented by the law on grassing of 8 June 1864.

21 Law of 4 April 1882 on the restoration and conservation of mountain plots.

22 Decoq, Kalaora and Vlassopoulos, *La Forêt salvatrice*.

23 Lefebvre et al., *Les Eaux et Forêts du XIIe au XXe siècle*, p. 582.

torrents des Hautes-Alpes, highly instrumental in spreading the image of the degradation of mountains among foresters, thus contrasted questionable climate action with incontestable pedological action.[24]

A progressive devaluation of the climatic action of forests was to be played out in this difference. Because the pedological effects were massive and sometimes evident, and because they were at the heart of the debates with Ponts et chaussées, over time the foresters' interventions put the emphasis on them at the expense of the climatic aspects.[25] The effects of trees on run-offs and springs stimulated research into watersheds, which proved pioneering in the history of modern hydrology.[26] Climatic influences seemed by comparison harder to grasp or to characterize. They would gradually become irrelevant to the arguments for protecting mountain forests and for reforestation. As foresters increasingly sought the origin of disasters beneath their feet, rather than in the sky, the link between forest and climate began to loosen.

This did not prevent some French foresters, like others elsewhere in Europe,[27] from researching the climatic action of trees. Thus, for more than ten years (1867–77), a professor at the forestry school of Nancy, Auguste Mathieu, carried out observations in woods, borders and fields to map this influence.[28] He concluded that in forests the temperature was lower, but more constant, and the air damper. But such work cut both

24 Alexandre Surell, Étude sur les torrents des Hautes-Alpes, Paris: Carilian-Gœury et V. Dalmont, 1841, pp. 145, 203. On Surell's importance in acknowledging the rise of erosion, see Corvol, *L'Homme aux bois*, pp. 272–9.

25 At the outset it was a question of responding to the challenges of Eugène Belgrand, 'De l'influence des forêts sur l'écoulement des eaux pluviales', *Annales des Ponts et chaussées*, 1st quarter 1854, pp. 1–27, and François Vallès, Études sur les inondations.

26 Vazken Andréassian, 'Waters and Forests: From Historical Controversy to Scientific Debate', *Journal of Hydrology*, no. 291, 2004, pp. 1–27 and 'Impact de l'évolution du couvert forestier sur le comportement hydrologique des bassins versants', ENGREF thesis, 2002, vol. 1, pp. 31–3. See also Brett M. Bennett and Gregory A. Barton, 'The Enduring Link between Forest Cover and Rainfall: A Historical Perspective on Science and Policy Discussions', *Forest Ecosystems*, vol. 5, 2018, article no. 5, pp. 1–9.

27 Ernst Ebermayer, *Die physikalischen Einwirkungen des Waldes auf Luft, Boden und seine klimatologische und hygienische Bedeutung*, Aschaffenburg: C. Krebs, 1873. For the same type of research in the Austro-Hungarian Empire, a little later and with results minimizing the possible climatic impact of forests, cf. Deborah R. Coen, *Climate in Motion*, pp. 269–70.

28 Auguste Mathieu, *Météorologie comparée agricole et forestière*, Paris: Imprimerie nationale, 1878. See also Léon Fautrat, *Observations météorologiques faites de 1874 à 1878*, Paris: Imprimerie nationale, 1878.

Archives départementales des Hautes-Alpes, collection of the RTM department, 21 Fi 1556.

Under the Third Republic, the forestry authorities increasingly invested in works to stabilize soil on mountains. Anti-erosion barrier in the stream of La Combe (Hautes-Alpes). Photograph by the Service de restauration des terrains de montagne, 1931.

ways, in that it revealed how difficult it was to deal with the issue. Simple though it seemed, the demonstration in fact raised enormous methodological problems. Depending on their position in the woods, rain gauges indicated different values; it was necessary to distinguish between rain and the condensation of water on plants; winds played an important role; comparisons were distorted by terrain and a thousand other factors. A major figure in fin-de-siècle climatology, Eduard Brückner, wrote in 1890: 'In general, all attempts to settle this question by comparative observation in forestry and non-forestry zones can be challenged on the grounds of uncontrollable factors that can cause local differences in precipitations, but which have nothing to do with the forest.'[29] When experimentally tested, the link between forests and climate seemed anything but obvious:

29 Eduard Brückner, 'Climate Change since 1790', in Nico Stehr and Hans von Storch, eds, *Eduard Brückner: The Sources and Consequences of Climate Changes and Climate Variability in Historical Times*, Dordrecht: Springer, 2000, pp. 77–191 (here, p. 99). In 1910 Gustave Huffel still had to respond in advance to methodological criticisms

so how, a fortiori, could one be sure about possible action from a distance? How could it be claimed that deforestation changed the climate of France, when it was already a struggle to demonstrate the micro-local action of trees?

The Slow Eclipse of the Forest Issue

As we have seen, essentially it was political struggles over forests that had made the climate a leading political issue since the Revolution. But this politicization would gradually sap the very credibility of the threat. One man had a particular responsibility here: François Arago. We have seen how, in the 1830s, he exploited the climate argument to persuade the National Assembly. He got his way, but the episode left its mark.

Anthropogenic climate change now appeared to be an ad hoc argument of a partisan kind, whose function was to influence political debate. The context reinforced this effect. Briefly close to the Thiers government, Arago shifted to the left from the late 1830s, expressing his republican opinions loud and clear. This activism made him a hero for some, but also clouded his credibility: had the politician prevailed over the scientist? Had he crossed a red line in unscrupulously exaggerating the risk of France's climatic degradation?

In parliamentary debates, the argument became the butt of ridicule: 'Refrain from doing an injustice when it is based on a mystery,'[30] mocked a deputy in the Chamber. In fact, article 219 was finally amended in 1859, in a way that exempted most woods from regulation.[31] This was a brilliant victory for liberals, who thereby won a battle for proprietary absolutism that had lasted for decades. It was also a mark of discredit for the idea of anthropogenic degradation of the French climate.

Sales of the national estate continued, however,[32] and the debates they

from those who disputed basic data on the action of trees: Huffel, Économie *forestière*, 2nd edn, Paris: Lucien Laveur, 1910, vol. 1, pp. 72–3.

30 *Journal de débats*, 21 June 1851.

31 A law passed on 18 June 1859 limited the interdiction to a closed list of situations: combating erosion, protecting springs and coasts, public health, and territorial defence. *Lois annotées, ou Lois, décrets, ordonnances, avis du Conseil d'État*, 1859, Paris: Bureaux de l'administration, 1860, p. 145.

32 30,990 hectares between 1860 and 1868. Laws of 20 July 1860, 19 May 1863, 8 June 1864, 18 July 1866 and 2 August 1868.

gave rise to did not drop the argument of climate change.[33] The press once more brandished the threat,[34] and supporters of the sales reached for mechanisms to get themselves off the hook – for example, a system of compensation whereby those who cleared land in the plains should replant an area equal or double in size in the Alps, the Pyrenees or the Corsican mountains.[35] Thus, the wooded surface area would not diminish and the climatic benefit of trees, should it exist, would be only the greater country-wide.

But these debates should not obscure a massive fact: the spectre of climate change caused by deforestation would soon recede in France. There were other reasons for this. First of all, from the 1860s the forestry estate entered a phase of uninterrupted expansion. As already mentioned, the activities of foresters in the mountains enabled the state to gain possession of more than a quarter-million hectares between 1860 and 1914. Added to this were the rich forests of Savoy, part of which joined the estate when the territory was integrated into France. And the expansion of the wooded estate was not restricted to mountainous zones. Thus, virtually 80,000 hectares of forest were nationalized by policies targeting regions of coastal dunes, like the Landes.[36] Furthermore, sales of national forest ceased completely from the 1870s to the present day.[37]

More generally, the spectre of a deforestation of the country receded under the impact of structural developments in the forestry economy. Private woodlands thrived, with high prices for the lumber siphoned off by industrialization and the enormous construction needs of railroads. Softwoods became a leading industrial resource alongside the traditional forests of Île-de-France or Burgundy. By the turn of the century, speculation no longer favoured clearance projects, but enterprises concerned with the purchase and afforestation of land.

---

33 Clavé, 'L'aliénation des forêts de l'État', pp. 197–214 and Études sur l'économie forestière, Paris: Guillaumin et Cie, 1862.

34 J. Rothschild, ed., *L'Aliénation des forêts de l'État devant l'opinion publique*, Paris: J. Rothschild, 1865.

35 *Revue des eaux et forêts*, 10 April 1865; *L'Opinion nationale*, 4 June 1865.

36 Henri Decencière Ferrandière, 'Origine des forêts domaniales', *Revue forestière française*, no. 4, April 1960, pp. 237–49.

37 More precisely, they now only concerned marginal surface areas, for the purposes of improvement or management at the micro-local level. 'Un siècle d'expansion des forêts françaises. De la statistique Daubrée à l'inventaire forestier de l'IGN', *L'Information forestière*, no. 31, May 2013, pp. 1–8.

Fears for France's forests were no longer so intense. And there was also less concern about the loss of specific ligneous resources. *La Couronne*, the first all-iron warship in the French navy, was launched in 1861: soon, lofty oak trees would no longer be vital resources for mastery of the seas. Thus it was that, under the Third Republic, the great battles over forest property and regulation, which had made anthropogenic climate change a burning issue since the Revolution, died down. Obviously, the idea did not disappear.[38] But, in France, it would never again exert the influence it enjoyed during three-quarters of a century. At the same time, a new land of conquest was opening up for foresters: the colonial Second Empire, from Algeria to French West Africa. As it gradually exited the stage in the metropolis, anthropogenic climate change was about to find new faces, new rationalities, and new roles far from the forests of the Alps and the benches of the Palais Bourbon.

The End of the Agricultural Ancien Régime

If the spectre of anthropogenic change receded at the century's close, this also stemmed from a reduced social vulnerability to the vagaries of the climate. For the first time in history, the chain linking bad weather, deficient harvests, food shortages, unrest, and violence was broken. Icy winters, dry springs and rotten summers were no longer synonymous with market chaos, panic and want.

The French diet improved spectacularly. The basic staples, consisting of cereals, became more abundant (+34 per cent for wheat between 1852 and 1881).[39] At the same time, meals diversified: more potatoes (+40 per cent), more meat (ditto), butter and eggs. Some products became more widespread: sugar, coffee, pasta, olive oil. Obviously, these developments differed between town and countryside, depending on regions and social classes. Still, they were widespread and durable in the second half of the century.

Food improved in quantity and quality, but also in terms of security of access. Historians date the last significant crises of subsistence in France

38 Cf., for example, Joachim-Mathieu Reynard, *L'Arbre*, Clermont-Ferrand: G. Mont-Louis, 1904, or André Jacquot, *La Forêt et son rôle dans la nature et la société*, Paris: Berger-Levrault, 1911. Decoq, Kalaora and Vlassopoulos, *La Forêt salvatrice.*

39 Average annual statistic per head. See Maurice Agulhon, Gabriel Désert and Robert Specklin, eds, *Histoire de la France rurale*, vol. 3, *Apogée et crise de la civilisation paysanne: 1789–1914* [1976], Paris: Seuil, 1987, p. 210.

to the middle of the century. In 1845-47 a mediocre harvest, followed by a rainy summer and then a dry spring, led to serious supply problems and social unrest throughout the country.[40] In 1853-57, the north of France suffered a deficit of cereal production that forced the state to assume supervision of the Paris market.[41] But these convulsions were the last. Naturally, neither high prices, nor poverty, nor economic slumps disappeared. Food was often costly. But the implacable connection between climatic vagaries and food shortages faded. The end of a long litany of shortages was clearly perceived by contemporaries.[42]

This break was bound up with a profound change in agricultural systems of production and exchange, often referred to as the end of an 'economic ancien régime'. It was first manifest in the rise in agricultural productivity, underway from the start of the century and accelerating after 1850. This increased productivity resulted from a transformation in the instruments of production (such as new models of plough) and the development of fertilizer.

Yet another factor was equally decisive, if not more so: the railway. The first tranche began to be constructed under the July Monarchy and was completed under the Second Empire. The network spread out from Paris to the major towns and cities. It was complemented, following the Freycinet plan of 1878, with a myriad of smaller lines like capillaries reaching to the heart of rural France. This rail infrastructure operated in synergy with a network of roads and minor paths that were the object of heavy investment in the second half of the nineteenth century.

Rail (coupled with canals, roads, and lanes) radically overhauled market structures. It created outlets for local products, even several hundred kilometres away. It made it possible to transport enormous quantities of grain; from the 1860s, one-third of their consumption in France transited by railway.[43] The network carried livestock over long distances and without the considerable losses of transport 'on the hoof'. The growth in flows was rapid and market integration boosted agricultural productivity.[44]

---

40 Nicolas Bourguignat, *Les Grains du désordre: l'État face aux violences frumentaires dans la première moitié du XIXe siècle*, Paris: Éditions de l'EHESS, 2001.

41 Laurent Herment, 'Les communautés rurales de Seine-et-Oise face à la crise frumentaire de 1853–1856', *Histoire et Mesure*, vol. 26, no. 1, 2011, pp. 187–220.

42 Roger Price, *The Modernization of Rural France: Communication Networks and Agricultural Market Structures in Nineteenth-Century France*, London/Melbourne/Sydney: Hutchinson, 1983, p. 197.

43 Ibid., p. 241.

44 With an index of 100 in 1913, cereals and flour sent by rail represented 0.61 in

Above all, the railway lessened people's vulnerability to climatic caprice. This occurred country-wide, between territories and between regions. In the event of bad harvests in the Beauce, for example, trains made it possible to source supplies from Brittany, where harvests had been good – as actually happened in 1866.[45] Abrupt spikes in the price of cereals, which could double or triple in a few days, became rarer. With these depreciations came a flattening of prices between localities, where once the differences could have been enormous over a mere 100 kilometres. The rail revolution operated in another way, too: it had a powerful psychological impact and diffused a new view of the market, which, foregrounding the ever-open possibility of importation, quelled panic effects. Panic frequently played a key role in subsistence crises, alongside opportunistic speculative behaviour.

Imports, too, took off with rail, combined with the development of maritime freight. The cost of maritime transport plummeted in the last third of the century: between 1879 and 1900, it was divided by seven between New York and Liverpool.[46] Russian cereals flooded into France via Odesa and Marseille; Le Havre received American wheat. France's cereal imports tripled between 1847–56 and 1887–96. Meanwhile, the population only grew by a little over 10 per cent.

As shortages became rarer, climate anxiety waned in France and other wealthy countries. Bad seasons no longer had the same political importance: the authorities were no longer on the look-out for them as the first link in a chain potentially leading to panics, riots and, sometimes, revolutions. Concerns about climate change receded, along with the worries about public order once spurred by sodden summers and punitive winters.

This was all the truer in that decreased social vulnerability to natural hazards was accompanied by increasingly favourable climatic circumstances. As contemporary historical climatology has shown, the 1860s marked the end of the Little Ice Age (LIA) that began in the 1300s.[47] This protracted chilly phase was succeeded by a series of rises in average temperatures, less frequent 'heavy winters', and the retreat of Alpine glaciers.[48]

1854, 44 in 1884, 51 in 1894, 71 in 1904; head of cattle, 39 in 1866, 43 in 1874, 69 in 1884, 73 in 1894, 92 in 1904: ibid., pp. 242, 246.

45 Ibid., pp. 242–3.

46 Wilhelm Abel, *Crises agraires en Europe: XIIIe–XXe siècle*, Paris: Flammarion, 1974, p. 38.

47 Le Roy Ladurie, *Histoire humaine et comparée du climat*, vol. 3, esp. pp. 13–55.

48 With the exception of a temporary reversal of the trend, approximately between 1950 and 1975. In recent decades, anthropogenic warming caused by CO_2 has abruptly accentuated the autogenous tendency to heat.

The first two decades of the post-LIA, the 1860s and 70s, were especially benign (Le Roy Ladurie speaks of the 1860s as an 'imperial feast'); they inaugurated the shrinking in the long run of the Alpine glaciers. Of course, there were exceptions, like the freezing winter of 1870–71 and the cold seasons of 1879. The latter case offers counterfactual evidence to Le Roy Ladurie: the price of wheat scarcely rose, whereas the historian of the 'economic ancien régime' would have anticipated a doubling or tripling. The railway, Russian and American wheats, and higher productivity all played their part.

Despite relapses, the trend was set: a favourable climate accompanied the eclipse, in public debate and government concerns alike, of the spectre of climate change in France.

The empire of climate retreated – for a while – in the face of the empire of technologies. Vulnerability to atmospheric fluctuation still existed, in sometimes unprecedented forms, but it no longer had the heft imparted by the threat of shortages. At the same time the state marshalled scientists and engineers to combat other risks. Systems for flood alerts benefited from the rapidity of the telegraph to warn towns and enable them to prepare themselves. At altitude, large-scale reforestation served to stabilize the soil, break up torrents, and regulate water behaviour. The upshot was ambiguous in terms of effectiveness. But in shifting attention from the sky to the ground over time, technological activity on the mountains helped repel climate anxieties. At the same time, the atmospheric sciences mutated. From the 1860s and 70s, scientists and states began to forecast 'what the weather is going to do' in order to keep the population informed.[49] As in the case of river surges, the telegraph played a key role in making it possible to monitor atmospheric phenomena in something close to real time. The Paris Observatory, and then the Central Meteorological Office, issued general forecasts and targeted alerts for sailors and farmers. In announcing coastal gales, hail, and storms twenty-four hours in advance, they too made the population less helpless before the vagaries of the sky. In this new world of rail tracks, locomotives, telegraph lines, corseted mountains and managed forests, people could finally dream of perhaps, one day, being completely free of the moods of the atmosphere.

49 Locher, *Le Savant et la tempête*; Fierro, *Histoire de la météorologie*.

14

The Enigmas of the Climatic Past

Throughout nineteenth-century Europe, debates raged between scientists about possible climate changes over decades, centuries, millennia and beyond. Anxieties about deforestation were an important catalyst in France, Russia, and later the Austro-Hungarian Empire and North America. But these fears did not exhaust the vast scientific and cultural horizon encompassed by the question of climate changes. With the expansion of Russian and US continental powers and of European colonial empires, archaeologists, naturalists and geographers encountered climates that appeared to have experienced a slow but steady natural alteration over the centuries. They wondered about the effect of these processes on the soil, the vegetation, and the societies of the past. The exploration of the world's soils and subsoils raised equally dizzying questions. Fossils, rocks, everything pointed to a very different planet in the remote geological past, exposed to torrid heat or freezing cold.

But science faltered: with no physical means of dating, with at best a few decades' retrospective observations, how was change to be grasped? How to get to the root causes? And how to understand the mechanisms in play, when it was already so difficult to identify what accounted for the regularity of climates, the recurrence of certain winds, certain cold or hot spells? Geological time would offer more certainty, but the climate of history remained an enigma which the atmospheric sciences side-lined for a while. Science would not have the last word on anthropogenic change. Worries about human climate action gradually faded in Europe and North

America at the end of the century, along with the political and material conditions that had determined their rise – they were not curtailed by the blade of learned words.

The Labyrinth of Change

Nineteenth-century wrangles about the historical evolution of climates involved experts of all kinds: geologists, geographers, botanists, agronomists, physicists, foresters, engineers, and archaeologists. There were plenty of viewpoints on the stability or, alternatively, the fluctuation of climates. Different causes of change were upheld: deforestation, solar influences, geological processes, volcanoes, modification of the Earth's orbit.[1] But no consensus could emerge in the absence of shared epistemological or professional norms, capable of converting disagreements into research programmes and shared analyses.

There was not even any consensus on the *evidence* for change. As we have seen, since the late eighteenth-century, historical references to extreme events (seasons, freezing of rivers, etc.) and the presence or absence of plants in certain regions had been cited as evidence of climate shifts. But, in the following century, this modus operandi was increasingly contested. Two processes contributed to that.

First of all, advances in plant geography, which was in full swing in the early decades of the nineteenth century. Among its findings was the quasi-impossibility of relying on *cultivated plants* to obtain a faithful description of the natural world. Cosseted and protected by gardeners, growers and landowners, these species were poor witnesses to the actual dynamics linking vegetation to the physical conditions of sites.[2] This largely disqualified them from informing any study of the climatic past, and yet the texts referred chiefly to them.

Such was the firm conclusion reached by the Danish botanist Joachim

1 Philipp N. Lehmann, 'Whither Climatology? Brückner's Climate Oscillations, Date Debates, and Dynamic Climatology', *History of Meteorology*, no. 7, 2015, pp. 49–70; Deborah R. Coen, 'The Advent of Climate Science', *Oxford Research Encyclopedia of Climate Science*, oxfordre.com/climatescience/view.

2 See, for example, the highly influential work by Augustin de Candolle, *Essai élémentaire de géographie botanique*, Strasbourg: F.G. Levrault, 1820, p. 9, or that by his son, Alphonse de Candolle, 'Distribution géographique des plantes alimentaires', *Bibliothèque universelle de Genève*, vol. 1, 1836, pp. 228–60.

Schouw.[3] His objective was to study climate changes using certain phenomena, such as the distribution of plants, as 'natural thermometers and hygrometers'.[4] But he found, and in concrete cases demonstrated, that if a plant was no longer grown in a place where it once had been, this did *not* licence the conclusion that there had been a change in climate. The same view was held by Charles Martins, one of the top European specialists in meteorology and plant geography. Reviewing a book on climate change in France, he drove the point home: the fact that, in the Middle Ages, vines were cultivated in Picardy proved nothing. In the past, people did not know how to preserve wine; transport was slow and costly, while agricultural land was readily available – so many factors that encouraged local cultivation and meant that 'the poor man himself had an interest in planting part of his inheritance in vines'.[5] The climate had nothing to do with it.

The rise of an ideal of precision in the philological and historical sciences tended in the same direction. As early as 1799, the American lexicographer Noah Webster was impugning evidence based on first-hand testimony, with its selective quotations, generalizations, and witnesses impressed by extreme but rare events. In 1843, one of his articles was republished to resounding effect.[6] At much the same time, the surgeon and meteorologist Samuel Forry published *The Climate of the United States*, sometimes regarded as the first substantial climatological description of that country.[7] In his book and in an article inspired by Webster, Forry tore to bits 'evidence' based on historical testimony.[8] Concurrently in France, the archivist Ludovic Lalanne also condemned those who tried their hand at it: how absurd to take accounts by medieval chroniclers

3 Joachim Schouw, 'On the Supposed Changes in the Meteorological Constitution of the Different Parts of the Earth during the Historical Period', *Edinburgh Journal of Science*, vol. 8, 1828, pp. 311–16. On Schouw, see Feldman, 'The Ancient Climate in the Eighteenth and Early Nineteenth Century', pp. 23–40 (esp. pp. 37–40).

4 Schouw, 'On the Supposed Changes', p. 311.

5 Charles Martins, 'Du climat de la France avant le XVIIe siècle', *Journal d'agriculture pratique*, May 1844, 2nd series, vol. 1, pp. 491–5.

6 Noah Webster, 'On the Supposed Change in the Temperature of Winter', *A Collection of Papers on Political, Literary, and Moral Subjects*, New York: Webster & Clark, 1843, pp. 119–62 (original edn 1810; a paper read in 1799). See Fleming, *Historical Perspectives on Climate Change*, pp. 47–50.

7 Samuel Forry, *The Climate of the United States and Its Endemic Influences*, New York: J. & H.G. Langley, 1842.

8 Samuel Forry, 'Researches in Elucidation of the Distribution of Heat over the Globe, and Especially of the Climatic Features Peculiar to the Region of the United States', *American Journal of Science*, no. 47, 1844, pp. 18–50, 221–41.

for observations of nature, when they were seeking to read the future in meteorological wonders.[9]

From the 1820s and 30s, the methods of historical climatology grew more refined. For example, Schouw used two plants whose studied sites were respectively at the northern and southern limits of viability: from this he deduced the temperature at the time their presence was recorded.[10]

It was also at this point that a crucial tool in contemporary historical climatology emerged: the utilization of 'harvest banns' (indicating the official start of the grape harvest) as a proxy.[11] This method, made famous among historians by the works of Emmanuel Le Roy Ladurie,[12] was proposed in the 1820s (cf. Chapter 11) and was first applied by an owner-harvester from Volnay, Denis Morelot.[13] But it was only diffused from the 1830s onwards, after an article by Jean-Étienne Noirot, a forestry specialist, listed a continuous series of *vendange* dates in Burgundy from the end of the fourteenth century. He concluded that there had been scant variation and therefore no change of climate.[14] For Noirot this was to strike a blow against forestry regulation, which was being debated in the Assembly at the time. His article made a big splash. In subsequent decades the method would become an important element in reflections on climates in the past, even if some pointed to its limitations given the influence of economic and cultural factors on the timing of harvests.[15] In spite of everything, it was used by all sides – those who thought that climates changed, those who believed in stability, and those who plumped for periodic alterations.

As Philipp Lehmann has shown, scientists who argued about past climates in the nineteenth century did not lack data.[16] They were positively

9 Ludovic Lalanne, 'Des changements dans le climat de la France, histoires de ses révolutions météorologiques, par le docteur Fuster', *Revue des Deux Mondes*, 30 April 1846, vol. 14, 1846, pp. 513–15.

10 For example, in the case of Palestine, the date palm and the vine. Schouw, 'On the Supposed Changes', pp. 315–17.

11 Because the start of the grape harvest depends on when fruit is ripe, and hence on summer temperatures, these dates can serve as a proxy.

12 Le Roy Ladurie, *Histoire du climat depuis l'an mil*, p. 8.

13 Denis Morelot, *Statistique oenologique de l'arrondissement de Beaune*, Paris: Mme Huzard, 1825, pp. 108–10 and *Statistique de la vigne pour la Côte-d'Or*, Dijon/Paris: Victor Lagier/Mme Huzard, 1831, pp. 214–18.

14 Jean-Étienne Noirot, 'Recherches sur les époques de la maturité des raisins', *Le Cultivateur*, 18 August 1836.

15 See, for example, Alexis Perrey, 'Sur la période des vendanges dans la Bourgogne', *Annuaire météorologique de France* for 1851, pp. 199–206.

16 Lehmann, 'Whither Climatology?'. This comprises field data, as in the works of the Austrian naturalist Anton Kerner von Marilaun, who mapped the distribution of

overwhelmed by the evidence dug up by the various parties, but without agreeing on its respective value, while coming under fire from ever-sharper methodological criticisms.

Such criticisms did not deter a teeming output on the subject. Given that it concerned anthropogenic change, of course, such output was continually boosted by the cycles of politicization that foregrounded the forestry question. As we have seen, this was the case in France between 1790 and 1870 – which explains, for example, why the naturalist and physicist Antoine-César Becquerel took an interest in it under the Second Empire. Questioned about deforestation in his capacity as an elected local official, and then witnessing the debates on forestry policy, he summarized half-a-century of works, and conducted experiments on the heat of trees and the temperature in woods to chart the effect of forests on the warmth of the air.[17]

In Russia, the debates crystallized in the years 1830–80 around the development of the steppe regions of the empire and the possibility of growing cereals there on a long-term basis.[18] At stake, in particular, was the potential climate change caused by deforestation. This prompted a wave of books on the subject, fuelled by the impact on senior officials and foresters of the French debates on climate and clearances in the 1830s.[19] The discussion was also informed by readings of Becquerel and George Perkins Marsh (cf. Chapter 15).[20] In Austria-Hungary, the issues of forestry regulation and the risks of flooding provoked similar debates between 1870 and 1900. This led the naturalist Josef Roman Lorenz to estimate the scale of the climatic effects of deforestation, specifying what level of government should be mobilized to confront them.[21]

---

wild plants to infer climatic alterations going back to the last ice age: Coen, *Climate in Motion*, pp. 274–311.

17 Antoine-César Becquerel, *Des climats et de l'influence qu'exercent les sols boisés et non boisés*, Paris: F. Didot frères, 1853, pp. 306–62, and *Mémoire sur les forêts et leur influence climatérique*, Paris: Firmin-Didot, 1865.

18 David Moon, 'The Debate over Climate Change in the Steppe Region in Nineteenth-Century Russia', *The Russian Review*, no. 69, 2010, pp. 251–75, and *The Plough That Broke the Steppes: Agriculture and Environment in Russia's Grasslands, 1700–1914*, Oxford University Press: Oxford 2013, esp. pp. 118–38.

19 Marina Loskutova, 'Quantifying Scarcity: Deforestation in the Upper Volga Region and Early Debates over Climate Change in Nineteenth-Century Russia', *European Review of History: Revue européenne d'histoire*, no. 27, 2020, pp. 253–72.

20 A summary of Becquerel's book appeared in Russian in 1854, only a year after its publication in Paris; as did G.P. Marsh's *Man and Nature* in 1866, two years after its appearance in English.

21 Coen, *Climate in Motion*, pp. 239–73.

These contributions fed into a field of study and reflection where at the same time possible autogenic alterations of climates were debated, as in the theories of Prince Pyotr Kropotkin on a change in the climate of Eurasia allegedly underway since the last ice age.[22] With this hypothesis he endeavoured to explain the geography of tundra, the steppes and dry prairies, as well as the archaeological enigma of so many ruins lost in deserts.[23] But, while he believed that the signs of this major desiccation had too often been taken as indices of human action, he did not hesitate to recommend – in accents that Buffon would not have rejected – planting wide bands of trees to counter creeping drought.[24]

The New Climate Sciences

Scientific approaches to the atmosphere underwent an overhaul after 1850. State meteorological bodies were created in most Western countries: Austria-Hungary (1851), Holland (1854), England (1854), Sweden (1859), Spain (1860), Italy (1863) United States (1870), France (1878). This wave of institutionalization often answered to the need to organize an operational weather forecasting service.[25] But these bodies also maintained climatological divisions, and observation networks often organized in tandem with the army, the navy, education and scientific establishments, technical administrations, and agricultural stations. The data generated was analysed and put into maps in order to characterize climatic regions and climate types.[26] The state climatology that emerged in the second half of the nineteenth century had no monopoly on climate discourse – it was equally a matter for geographers, naturalists or foresters, each with their own methods and research agendas. But its

22 On Kropotkin and his theory, see Mike Davis, 'The Coming Desert: Kropotkin, Mars and the Pulse of Asia', *New Left Review*, no. 97, 2016, pp. 23–43.

23 Cf., for example, Prince Kropotkin, 'The Desiccation of Eur-Asia', *Geographical Journal*, vol. 23, no. 6, 1904, pp. 722–34.

24 Ibid., p. 734.

25 Katharine Anderson, *Predicting the Weather: Victorians and the Science of Meteorology*, Chicago: University of Chicago Press, 2005; Locher, *Le Savant et la Tempête*; Fleming, *Meteorology in America, 1800–1870*, Baltimore: Johns Hopkins University Press, 1990.

26 Paul Edwards, *A Vast Machine: Computer Model, Climate Data, and the Politics of Global Warming*, Cambridge, MA: MIT Press, 2010, pp. 61–81.

institutional power and epistemic legitimacy made it a major protagonist from the 1860s and 70s onwards.

In some countries, state climatology developed as a radically descriptive scientific practice, aiming to document average atmospheric conditions regarded as quasi-stable on the scale of the century or the decade. This was true of the United States, where Cleveland Abbe, the first chief meteorologist of the Weather Bureau, wrote in 1889:

> No important climate change has yet been demonstrated since human history began. . . . The true problem for the climatologist to settle during the present century is not whether the climate has lately changed, but what our present climate is, what its well-defined features are, *and how they can be most clearly expressed in numbers.*[27]

One of his successors, Willis Moore, chief meteorologist in the 1910s, did not believe in rapid and/or man-made climate change either. He was scathing about meteorological measurements in forests, such as those made by Mathieu and Fautrat in France (Chapter 13), which claimed to demonstrate an effect – a local one – of trees on precipitations. He declared that no datum from the Weather Bureau indicated any change in US rainfall.[28] In France, the situation was the same: the Central Meteorological Office concentrated its efforts on producing precise observations, in order to describe regional climates and encourage research in forecasting meteorology. The issue of climate change was not on the agenda. Alfred Angot, the office's chief climatologist, regarded climates as generally stable and was sceptical about the existence of human climate action. He stressed that humanity would have to transform vastly large surface areas for its impact to be significant.[29]

27 Quoted by Fleming, *Historical Perspectives on Climate Change*, p. 53. On Abbe, see Edmund P. Willis and William H. Hooke, 'Cleveland Abbe and American Meteorology, 1871–1901', *Bulletin of the American Meteorological Society*, vol. 87, no. 3, 2006, pp. 315–26. For a recent essay in interpretation of the 'stable' climate paradigm promoted by state climatology in the US, see Zeke Baker, 'Agricultural Capitalism, Climatology and the "Stabilization" of Climate in the USA, 1850–1920', *British Journal of Sociology*, DOI: 10.11 11/1468-4446.12762 (accessed 29 June 2020).

28 By contrast, Moore believed in autogenic climatic alterations on the scale of a millennium or even a century. Willis Moore, *A Report on the Influence of Forests on Climate and on Floods*, Washington: Government Printing Office, 1910.

29 Alfred Angot, *Traité élémentaire de météorologie*, Paris: Gauthier-Villars, 1899, pp. 409–10.

Not all national meteorological bodies confined themselves to purely descriptive approaches. As Deborah Coen has shown, in the last third of the century the Austro-Hungarian ZAMG[30] formed the crucible of a new way of understanding climatic realities.[31] This 'dynamic climatology' sought to apply the physics of fluids and heat to the atmosphere and its interactions with the continents, oceans, reliefs and land surfaces, in order to explain climatic regularities in the light of global atmospheric dynamics. But it was not until the 1920s that this approach, combined with input from statistical physics, was applied to the study of climate stability and climate change.[32] And it seems this was done without further scrutiny of the possible climatic impact of deforestation.[33] The rise of dynamic meteorology at the start of the twentieth century was no more propitious to analyses of climate change.[34] In taking the physical equations governing the atmosphere as its starting-point, this approach came up against a sizeable problem: the impossibility of solving, in general, these equations, which made it necessary to employ approximate methods and, later, computer simulation. In the meantime, science remained very poorly equipped for thinking about possible anthropogenic climate changes.

In 1890, the German climatologist and geographer Eduard Brückner published a vast work on climate changes, in which he drew up an overall balance sheet of the existing research.[35] The material he read on the climate in history left him feeling confused and disorientated:

> We have arrived at the end of our overview. We have surveyed a labyrinth, but without an Ariadne's thread. Again and again we have encountered the same insoluble and irreconcilable contradictions. This

30 Zentralanstalt für Meteorologie und Geodynamik.

31 Coen, *Climate in Motion*, esp. pp. 171–236 and 'The Advent of Climate Science'.

32 With the works of the Italian-Austrian Albert Defant, studied by Coen in *Climate in Motion*, pp. 228–33.

33 And this even though this type of application had been encouraged by Wilhelm Schmidt, a pioneer of such approaches: Coen, *Climate in Motion*, pp. 226–7.

34 On the development of a mathematized theory of atmospheric processes, see notably Robert M. Friedman, *Appropriating the Weather: Vilhelm Bjerknes and the Construction of a Modern Meteorology*, Ithaca: Cornell University Press, 1989, and Edwards, *A Vast Machine*.

35 Eduard Brückner, *Klimaschwankungen seit 1700. Nebst Bemerkungen über die Klimaschwankungen der Diluvialzeit*, Vienna and Olmütz: Hölzel, 1890. We have consulted the English translation in Stehr and von Storch, eds, *Eduard Brückner*, pp. 77–125. On Brückner, see ibid. and Lehmann, 'Whither Climatology?'.

> is virtually a mystery for thought: that serious scientists have diagnosed mutually exclusive climate changes for the same countries. It is just as disconcerting to note that the forest is constantly designated as the scapegoat for a whole series of often contradictory changes. Something rather striking emerges from this: the absence of any progress towards a solution.[36]

Brückner had his own agenda: based on an enormous mass of data on harvest dates, glaciers and rains, he mooted the existence of *cyclical* variations in climates, natural in origin and lasting for a period of thirty-five years.

Here, though, he was not merely arguing on his own behalf. He also captured something of the discredit into which the theme of historical climate change had now fallen among some scientists: 'In this confusion of contradictory opinions, which are often unfounded, it does not seem astonishing that at present the very fact of dealing with the question of climate change, and still more adding a new hypothesis to the existing ones, virtually violates the *modus vivendi* between meteorologists.'[37]

Supporters of state climatology and of new dynamic approaches to the atmosphere would tend to set aside this subject, which increasingly seemed like an aporia, an insoluble problem that doomed those who engaged with it to a labyrinth of hypotheses and unanswered questions. Anxious to display the scientific character of their approaches, these climatologists bet on stability or – just as often – stated that they were not in a position to decide. The most influential work in fin-de-siècle climatic approaches, the *Handbuch der Klimatologie* by the Austrian Julius Ferdinand van Hann, thus explained that no solid proof existed either to demonstrate *or to refute* 'progressive changes' to the climate caused by man or perceptible at his level.[38] And this following a century of research on the subject.

Brückner's theory further justified this lack of interest by opening up a path to conceiving change, but within stability, and without man playing any part in it. His theory of cycles was popular for reinterpreting the available data under the assumption of oscillations around a fixed point. Through it, Brückner also proposed an abrasive re-reading of theories

36 Stehr and von Storch, eds, *Eduard Brücker*, pp. 115–16.

37 Ibid., p. 78.

38 Julius Ferdinand von Hann, *Handbook of Climatology* (1883), New York/London: Macmillan Company, 1903, pp. 401–2 (translation of vol. 1 of the 1897 edition).

of change. According to him, a statistical inventory of these, grouped by decade since 1790, proved that all their authors were wrong, but also, in a sense, right. In their time, he explained, they were describing the trend then underway in the periodic curve of thirty-five years. What he called the 'labyrinth of hypotheses and opinions on climate change' found its order and explanation here: the psychological effect of a variation around a fixed state, generating the (mistaken) impression of a trend. Climate change as a misstep of reason.

The Furnace of the Carboniferous

In response to the proliferation and impasse of debates on the recent development of climates, a new climatic history of geological time emerged in the nineteenth century.

Initially, it took the form of a radically new reading of Buffon's cooling and its effects on terrestrial climates. We owe it to a scientist, Jean-Baptiste Joseph Fourier, often cited as the 'founder' of theories of the greenhouse effect. On this point, the reality is more subtle and more interesting (cf. Chapter 16). He is also known as the person who first formulated the laws governing heat.[39] But his interests were not limited to theoretical physics. In 1809 he applied his equations to the distribution of terrestrial temperatures caused by solar rays.[40] Then, for ten years, he dropped the subject.

But, with the glacial winter of 1819–20, Buffon's thesis of a cooling of the globe – and of climates – once again aroused public interest. Fourier reacted with a project: to subject the 'hypothesis of an internal and central heat' to an 'exact analysis based on the knowledge of the mathematical laws of the propagation of heat'.[41] Combining calculations and observations made in grottos and mines, he arrived at an order of magnitude of

39 Ivo Grattan-Guinness, *Joseph Fourier. 1768–1830*, Cambridge, MA/London: MIT Press, 1972; John Herivel, *Joseph Fourier: The Man and the Physicist*, Oxford: Clarendon Press, 1975; Jean Dhombres and Jean-Bernard Robert, *Fourier, créateur de la physique mathématique*, Paris: Belin, 1998.

40 This work was published much later: Jean-Baptiste Joseph Fourier, 'Théorie du mouvement de la chaleur dans les corps solides (suite)', *Mémoires de l'Académie des sciences*, vol. 5 [1821–22], 1826, pp. 153–246.

41 Jean-Baptiste Joseph Fourier, 'Extrait d'un Mémoire sur le refroidissement séculaire du globe terrestre', *Annales de chimie et de physique*, vol. 13, 1820, pp. 418–38 (here, p. 420).

the excess temperature created on the surface by heat from the interior of the Earth. This excess was minimal, around three-hundredths of a degree. As for the fall in temperature experienced by humanity, it could not have exceeded three-*thousandths* of a degree! Fourier did not deny that the Earth was growing colder – but over a very long term in which man did not feature. His works had great impact and were very widely taken up. Fourier, said Arago in delivering his funeral eulogy, 'substituted evidence and proofs' for the old conceptions about the planet: he had shown that the widely dreaded freeze was 'purely a dream'.[42] This was the *coup de grâce* for Buffon's theory of a spontaneous cooling of climates that might be perceived on the scale of human history.

In those same years, science advanced on other fronts. Scientists were fascinated by the foreign and exotic plants that had left traces in European subsoils.[43] Palaeobotany emerged in the wake of palaeontology. In the 1820s and 30s, the study of ancient animals developed throughout Europe under the impetus of scientists like William Buckland, Henry de La Beche and Georges Cuvier. Their method? Mapping groups of extinct animals whose remains were found in subsoils by relating them to the rocks from which they came.

Fossil botany was much more difficult. The reproductive organs (flowers, fruits, seeds), central to Linnean classification had only rarely been fossilized. The plants themselves appeared in a seriously diminished form. The challenge was taken up at the start of the 1820s by a young scientist, Adolphe Brongniart. In 1822, he showed that certain fossil plants identified in the subsoils of Europe corresponded to vegetation of a tropical kind, ferns in particular.[44] These hot-country plants had been discovered in Carboniferous strata, layers that commanded attention because they harboured coal seams.

This indirect evidence confirmed what had been intuited since the closing decades of the eighteenth century (Chapter 3): in a remote geological past, Europe had been warmer than it now was. In 1828 Brongniart

42 François Arago, *Éloge historique de Joseph Fourier*, Paris: Firmin-Didot, 1833, pp. 51, 54.

43 On the emergence of palaeobotany and theories of the Carboniferous, see Martin J.S. Rudwick, *Worlds Before Adam: The Reconstruction of Geohistory in the Age of Reform*, Chicago: Chicago University Press, 2008, pp. 55–7, 167–72.

44 Adolphe Brongniart, 'Sur la classification et la distribution des végétaux fossiles en général, et sur ceux des terrains de sédiment supérieur en particulier', *Mémoires du Muséum d'histoire naturelle*, vol. 8, 1822, pp. 203–40, 297–348.

went further, proposing a general evolutionary schema linking the history of the Earth, of the climate and of plants.[45] According to him, the early Earth was very hot and its atmosphere saturated with carbon dioxide. Thanks to this copious gas, primitive vegetation would have been able to develop, even in the absence of fertile soil, and then be deposited in the ground in the form of coal. The Earth's atmosphere had thus been purged of its carbon dioxide, while cooling down over time in keeping with the process described by Buffon and Fourier.[46] A gradual transition had been effected from the lush world of the Carboniferous to present-day nature.

Man was excluded from the tropical world thus revealed. Advances in stratigraphy (the study of geological layers) provided, at the same time, outlines of chronological scale for past life forms. Human remains only appeared near the topmost surface: human history was but a tiny fraction of the history of nature.

The world of the Carboniferous, by comparison, seemed infinitely distant. Yet, as Fredrik Albritton Jonsson has shown, this did not prevent dizzying comparisons. In Victorian England, the luxuriance of the Carboniferous period was interpreted by scientists, and known to the public at large, as the means whereby Providence had made it possible for modern societies to develop thanks to coal and a breathable atmosphere.[47] Neither mankind nor industrial society could have existed without the steamy climates of the deep past.

Entering the Holocene

But the study of very ancient climes in these years did not only invoke heat and luxuriance. There had also been a planet gripped by ice and cold.[48] As we have seen, in Switzerland in 1820–21, the forest administrator

45 Adolphe Brongniart, *Prodrome d'une histoire des végétaux fossiles*, Paris: Levrault, 1828, pp. 183–88, 217–33.

46 Brongniart also thought that more abrupt bursts of change might have been superimposed, like those envisaged by Cuvier: Rudwick, *Worlds Before Adam*, p. 172.

47 Fredrik Albritton Jonsson, 'Holocene by gaslight', talk in Paris (2015) kindly communicated by the author, and Fredrik Albritton Jonsson, 'Abundance and Scarcity in Geological Time, 1784–1844', in Sophie Smith and Katrina Forrester, eds, *Nature, Action and the Future: Political Thought and the Environment*, Cambridge: University of Cambridge Press, pp. 70–93.

48 There is an abundant literature on the history of the theory of ice ages. We may cite John Imbrie and Katherine Palmer Imbrie, *Ice Ages: Solving the Mystery*, Cambridge,

Karl Kasthofer and the engineer Ignace Venetz responded to the call of the Helvetic Society of Natural Sciences with two papers on the likelihood of change in the climate of the Alps. Summoned to pronounce on the threat of the cooling predicted by Buffon, both minimized – not to say excluded – the possibility. Instead, they stressed the existence of enormous glaciers that used to cover the Swiss mountains in very remote times.

Their papers were not only designed to reassure the Swiss authorities. They also intervened in an ongoing debate among scientists about 'erratic boulders'. The latter were enormous rocks, different in composition from the surrounding terrain and bearing marks of movement on their surface. They had been identified in the Alps and also in Germany and Scandinavia. There were plenty of hypotheses for their presence, from a mega-tsunami to a large-scale flood; none enjoyed unanimity.

The belated publication of Venetz's memoir in 1829 and 1833 changed the picture. In effect, he offered a new explanation: erratic boulders, he suggested, had been carried along by glaciers that had left them *in situ* when they receded. The idea was taken up by the German-Swiss geologist Jean de Charpentier. In 1835, he published an article describing a 'monster glacier' that covered the upper valley of the Rhône in the distant past. Then another Swiss, Louis Agassiz, pushed the hypothesis to the extreme and made the crucial leap. In Neuchâtel in 1837, he presented the first theory of ice ages.[49] Agassiz no longer referred to outsize glaciers; for him the whole of Europe was once covered by a crust of ice. This crust, he said, had melted as the history of the Earth unfolded, to leave, first, Charpentier's giant glacier, and then just a few residues. Despite their massive size, present-day glaciers were merely the derisory remains of a vast world of ice.

But how could it have been so much colder in the past, when everyone regarded global cooling – over geological time spans – as an undeniable reality? It was necessary, Agassiz argued, to accept that cooling occurred while itself undergoing variations that transported the Earth into states of extreme cold at times. And with the aid of a few fine-tuned typographical

MA: Harvard University Press, 1986; Rudwick, *Worlds before Adam*; Tobias Krüger, *Discovering the Ice Ages: International Reception and Consequences for a Historical Understanding of Climate*, Leiden: Brill, 2013.

49 Louis Agassiz, 'Discours prononcé à l'ouverture des séances de la Société helvétique des sciences naturelles, à Neuchâtel le 24 juillet 1837', *Actes de la Société helvétique des sciences naturelles*, 22nd session (1837), pp. v–xxxii.

signs, he proposed a 'curve' of the globe's temperature on the scale of geological time:

The subject of lively debates from 1840–60, Agassiz's theory finally carried the day. Scientists now accepted that in the distant geological past, the north of the Northern Hemisphere was a very cold zone covered in ice.

The causes of ice ages were initially a mystery. The French mathematician Joseph-Alphonse Adhémar was the first to suggest an astronomical cause in his 1842 book, *Révolutions de la mer*.[50] He thought the great crust of ice described by Agassiz might be an extension of the North Pole's ice cap, whose disappearance was to be explained by the precession of the equinoxes – the slow change in direction of the Earth's axis. In 1864, in a momentous article, the Scot James Croll drew on Adhémar to present a much more ambitious and complete theory of geological climate changes.[51] Croll considered not only the precession but also the evolution of the Earth's orbit, the weight of mantles of ice, ocean currents, and the circulation of the atmosphere in order to explain the alternation of glacial phases and warmer sequences. In the twentieth century this theory was further perfected by the Serb Milutin Milanković. It forms the basis of our current knowledge of ice ages.

What place was there for mankind in this picture? Thanks to the advances of physics, stratigraphy, palaeontology and the theory of ice ages, it was known by the mid-nineteenth century that climates had a long, complex history from which human beings were largely excluded – as they were from most of the history of living things. It was to make room for humanity in the vast chronology of life that the palaeontologist Paul Gervais coined the term 'Holocene' in 1850.[52] Defining as 'holocene environments' those where, in addition to animal remains, traces of a human presence could be detected, he called the corresponding stretch of time the 'Holocene age' (or 'anthropogenic age'), while ascribing only minor

---

50 Joseph-Alphonse Adhémar, *Révolutions de la mer. Déluges périodiques*, Paris: Carilian-Gœury and V. Dalmont, 1842.

51 James Croll, 'On the Physical Cause of the Change of Climate during Geological Epochs', *Philosophical Magazine*, no. 28, 1864, pp. 121–37. On Croll and his theory, see James R. Fleming, 'James Croll in Context: The Encounter between Climate Dynamics and Geology in the Second Half of the Nineteenth Century', *History of Meteorology*, no. 3, 2006, pp. 43–53.

52 Paul Gervais, 'Sur la répartition des mammifères fossiles entre les différents étages tertiaires qui concourent à former le sol de la France', *Mémoires de la section des sciences de l'Académie des sciences et lettres de Montpellier*, 1850, pp. 399–413 (here, p. 413).

importance to it in terms of the long history of life on Earth.[53] The term came into use when, in the final quarter of the century, the need was felt for a standardized international geological chronology. It was promoted by the Swiss Eugène Renevier, a little-known figure whose role in establishing modern geochronology was central.[54]

Also at this time, the Scot geologist James Geikie tackled one of the great mysteries of prehistory: the chronology and causes of the transition from the Palaeolithic to the Neolithic eras. For him it was the emergence from the last ice age that underlaid this dramatic change, the origin of modern societies. Fredrik Albritton Jonsson has highlighted the influence of this thesis, which became established at the end of the nineteenth century even though there is still no consensus as to a precise dating.[55]

At the outset there were human beings crushed by the cold, sheltering in caves, and hunting reindeer and woolly rhinoceros. Then, a major return of warmer weather facilitated the expansion of agriculture, towns, and complex societies. By making the start of the Holocene and of the Neolithic age coincide in his chronologies, Renevier completed the picture: the favourable climate of the Holocene as the condition of possibility for all major human civilizations, from Ancient Egypt to the Victorian Empire.

Thus, at the end of the nineteenth century, humanity discovered a climatic homeland, mild and reassuring compared with the sweltering heat of the Carboniferous and the great cold of the ice ages. With the category of the Holocene, scientists fixed a scale of change that relocated the human history of the climate within a much vaster chronology, of which it was no more than a minuscule fraction. The great battles over historical changes in climates did not come to an end. But the new scientific approaches to the atmosphere tended to neglect these issues and some scientists now regarded them as an aporia, a mystery to be put to one side, the better to progress. Others saw a climate that was transforming itself, but only in accordance with oscillations that would return it, after

53 Paul Gervais, 'Nouvelles remarques concernant la répartition des mammifères entre les différents étages tertiaires (2e partie)', *Comptes rendus de l'Académie des sciences de Paris*, no. 34, 1852, pp. 520–4 ('Age holocène', p. 523).

54 Eugène Renevier, *Chronographe géologique*, Lausanne: G. Bridel, 1897.

55 Jonsson, 'Holocene by Gaslight'; James Giekie, *The Great Ice Age, and Its Relation to the Antiquity of Man*, New York: Appleton & Co., 1874. For another example of a theory linking climate dynamics and human history to prehistory, see Henri Le Hon, *L'Homme fossile en Europe: son industrie, ses mœurs, ses œuvres d'art*, Paris: C. Reinwald, 1867.

a gap of centuries or decades, to the starting point. In his great book on the Mediterranean, Fernand Braudel would invite historians to investigate these climatic cycles of cold and heat, aridity and humidity, which he likened to those of economic life.[56] Nevertheless, the Holocene remained a battlefield and a mystery to be unravelled for all those who, in a ruin or an old citation, searched for a clue to a physical transformation that does not amount to the return of the same.

Across Europe, the spectre of climatic degradation caused by man's destruction of forests gradually faded away. The eclipse of the great nineteenth-century disputes over forestry regulation and state ownership, the end of the pre-eminence of wood as a source of energy, the impact of technological modernity on climatic vulnerability and agricultural productivity, plus the absence of convincing scientific evidence after decades of research – everything combined to roll back the threat. Science had not refuted the paradigm of human climate action, an idea that still pervaded the empires at this time. But, at the dawn of the twentieth century, the threat no longer possessed the social and political charge that had placed it, in France and then throughout Europe, at the heart of the major upheavals of the age.

---

56 Fernand Braudel, *La Méditerranée et le monde méditerranéen à l'époque de Philippe II* [1949], Paris: Armand Colin, 1990, vol. 1, pp. 245–9. In connection with climatic cycles, he cites Emmanuel de Martonne, *La France. Première partie France physique*, Paris: Armand Colin, 1942, p. 140, a work itself inspired by Brückner.

15

Restoring the World, Governing Empires

In the nineteenth and twentieth centuries, the British and French colonial empires extended over much of the known world. Whole continents of jungles, deserts, savannas, and cultivated areas were ruled by colonial elites desirous of controlling them so as to 'develop' them. We have seen that the belief in Western superiority based on mastery of nature was an old one, and in the Americas of the sixteenth to the eighteenth centuries it already underpinned the project of *improving the climate*, civilizing its features through deforestation and agriculture.

But the empires of the nineteenth and twentieth centuries added a new dimension: colonized peoples were now accused of having *damaged* their own climates. The colonizers denounced the impact of native agricultural and forestry practices on rains, winds, and temperatures. The Europeans then assigned to themselves the mission of *protecting* and *restoring* climates. In the imperial context, the paradigm of climatic degradation long remained highly influential – something much better known today, thanks to the work of historians such as Diana K. Davis, Caroline Ford, or Gregory Barton.

This influence was all the greater in that it allowed the conquerors to disclaim responsibility for enormous, long under-estimated tragedies: the terrible famines that struck India, the Maghreb, and French West Africa in those years. These colonial catastrophes arose from the climatic vulnerability of regions whose means of subsistence were destabilized by conquest and their integration into the capitalist world-system.[1] For the colonial powers, they were at once affronts to their claim to be developing the territories, and scientific puzzles to be elucidated. Here, narratives of climate degradation furnished them with the perfect interpretative grid: notwithstanding appearances to the contrary, famine was not a sign of their own bad management of populations and territories but pointed to the destruction of the environment by the colonized themselves, requiring, paradoxically, a reinforcement of imperial power.

The Arabs and the Climate

In the French Empire, the Algerian Arab populations were the first to be targeted. The history of the 'environmental myth' of the colonial Maghreb has been analysed in detail by Diana K. Davis, on whose work this section is largely based.[2]

The rhetoric of climatic culpability and decline emerged in the 1840s, initially from the pen of Jean-Jacques Baude, a politician and senior civil servant.[3] Baude was a figure in the Orleanist party, tasked by King Louis-Philippe in 1836 with a fact-finding tour of Algeria. He derived a book from this experience which, for the first time, deployed the grand narrative of decline. The Algerian coastline, he explained, was falling prey to the destruction of its forests through felling, grazing, and the bush fires set by the natives. The stakes, Baude asserted, were immense because the trees attracted rain and nourished water resources. France must act to protect and restore these forests. This, Baude wrote, would be 'a powerful way of improving the climate'.[4]

All this echoes the debates agitating the metropolis at the time. Baude, who was a deputy, took part in the parliamentary discussions on the authorization of clearances. He witnessed Arago in the National Assembly warning against the climatic dangers of deforestation. He himself was an

1 Mike Davis, *Late Victorian Holocausts: El Niño Famines and the Making of the Third World*, London and New York: Verso, 2001; Brock Cutler, 'Evoking the State Environmental Policy and Colonial Disaster in Algeria, 1840–1870', PhD thesis, University of California-Irvine, 2011.

2 Diana K. Davis, *Resurrecting the Granary of Rome: Environmental History and French Colonial Expansion in North Africa*, Athens: Ohio University Press, 2007.

3 Baron Jean-Jacques Baude, *L'Algérie*, Paris: Arthus Bertrand, 1841, vol. 1.

4 Ibid., p. 113.

ardent supporter of forestry regulations, and lamented that it was such a struggle to impose them in France. In this regard, Algeria offered fabulous opportunities.[5] Over there, he rejoiced, there was (almost) nothing to impede the action of the state: neither private property nor the political leverage of the inhabitants.

Baude's voice was not an isolated one. The monumental *Exploration scientifique de l'Algérie* published in the 1840s–1860s also contributed to the formation of an anti-Arab, declinist grand narrative. It gave it historical depth: formerly vast, the Algerian forest had supposedly receded under the axe of Arab conquerors, who overran the area in the twelfth century and imposed their way of life.[6] Cue scarce rains, distorted winds and dried-up springs.

Here it was – the broad outline of a narrative that would be wheeled out time and again, from the early stages of the conquest to just after the Second World War.[7] It was ubiquitous and consensual in Algerian colonial society from the late 1860s.

This story drew on a whole imaginary of collapse, fed by reflections on the fall of empires, the Orient, and the role of climate in the end of civilizations. From the end of the eighteenth century, as we have seen, people were haunted by visions of ruined cities buried in the sands, the vestiges of the suicide by climate of entire societies. They served to raise the alarm as regards France. Bernardin de Saint-Pierre, Rougier de La Bergerie, Chateaubriand and Mirbel pointed to Persia, Egypt, Arabia and Judea, where deforestation had parched everything.[8] Rauch made it the third act of his grand narrative on climate and civilization (Chapter 10): in it we see Nineveh and Babylon, Sumer and Asia Minor, those cradles of civilization, perishing from having felled their trees and changed their climates.[9] These dystopian visions were projected onto the Roman remains dotting Algeria: these ruins were interpreted as all that survived of a formerly prosperous, now barren country.

5 Ibid., p. 116.

6 Jean-André Napoléon Périer, 'De l'hygiène de l'Algérie', *Exploration scientifique de l'Algérie. Sciences médicales*, vol. 1, Paris: Imprimerie royale, vol. 1.

7 See Davis, *Resurrecting the Granary of Rome*, which follows its thread and provides a comprehensive review of the literature.

8 Bernardin de Saint-Pierre, *Études de la nature*, vol. 2, p. 324; *Archives parlementaires*, 21 March 1817; Mirbel, Éléments de physiologie végétale et de botanique, p. 452. See also Baudrillart, *Traité général des eaux et forêts*, vol. 1, p. 27.

9 Rauch, *Harmonie hydrovégétale et météorologique*, vol. 1, p. 45 and *Annales européennes*, vol. 1, 1821, pp. 105–6, 110.

This discourse on climate change exhibited a powerful ethno-racial dimension in Algeria. The Arab 'race' was condemned for its inherently destructive character.[10] Three other racist discourses arose around this anti-Arab eco-racism, confirming it by way of contrast.[11] First, one that assimilated the French to the race of Romans who populated the Algeria of Antiquity. Thanks to this fiction, the Arabs could be painted as invaders! The Roman ruins of the Maghreb were interpreted here not only in the light of climate collapsology, but also as traces of a moral, productive, orderly civilization whose revival in a modern form was the mission of the French empire.

The second racial vision was presented as a hope: seeing the emergence in Algeria of a new 'Latin race', strong and fertile, issued from the mixing of colonists from France but also Italy and Spain.[12] To let the Maghreb dry out further would be to compromise the vitality of the European races destined to develop it.

Finally, a third racial theory pitted 'Arabs' against 'Kabyles'. Colonial discourse described the latter as the remnants of the Vandal (hence European) populations forced by the Arab invasions to take refuge on the margins. The contrast between the figure of the 'Arab' and the 'Kabyle' is a familiar topos, a sub-product of French domination in North Africa. It has an ecological plank: the Arab was allegedly the enemy of the tree and hence of the climate; the Kabyle knew how to manage his environment, as testified by the forests of that province.

Eco-racism was very widespread in Algeria until the late twentieth century. Thunderous on the spot, it also found echoes in the Hexagon. Thus, in 1850, a writer in *Annales forestières* deplored the fall in the funding of Algerian forestry, before denouncing those 'companies of Arabs, of a different species' – speculators – who in the metropolis aspired to get their

10 In *Resurrecting the Granary of Rome*, Diana Davis stresses the indictment of the nomadic way of life, which is linked in colonial discourse to a disdain for property and 'development'.

11 Caroline Ford, 'Reforestation, Landscape Conservation, and the Anxieties of Empire in French Colonial Algeria', *American Historical Review*, vol. 113, no. 2, 2008, pp. 341–62; Diana K. Davis, 'Restoring Roman Nature: North African Environmental History', in Davis and Edmund Burke III, eds, *Environmental Imaginaries of the Middle East and North Africa*, Athens: Ohio University Press, 2011, pp. 60–87; Patricia M.E. Lorcin, 'Rome and France in Africa: Recovering Algeria's Latin Past', *French Historical Studies*, vol. 25, no. 2, 2002, pp. 295–329.

12 Ford, 'Reforestation, Landscape Conservation'.

hands on the mainland mountain ranges, taking advantage of the sales of the national estate.[13]

As Diana K. Davis has emphasized, the grand narrative of climatic decline was closely linked to the expropriation of Algerian community forests – a process that became massive from 1850, with nationalization and a drastic restriction of user rights (felling, grazing, controlled fires).[14] It served to justify this policy by constructing a perceptual framework in which colonial coercion was established as legitimate for the sake of conserving nature and resources. This manoeuvre became flagrant after Napoleon III attempted to limit land dispossessions in the 1860s. The counterattack of the colonists' party? Propagating the argument of climatic decline. The natives did not deserve their forests. Indeed, proclaimed one colonist, 'France itself, so prosperous, would soon be a desert if it was in Arab hands.'[15]

The colonists' party also exploited a devastating catastrophe within the country. In 1866–68, a drought, compounded by a plague of insects and an earthquake, caused the deaths of around 800,000 people – one-fifth of the population.[16] For the supporters of the narrative of degradation, these corpses made ideal culprits. In a report of late 1867 to the colony's governor-general, the agronomist Robiou de La Tréhonnais devoted a long discussion to climate change and the Arabs' responsibility.[17] Others, like the journalist Henri Verne, perceived an even more immediate cause: the fires lit by Arabs in 1863–65, which appeared as a form of resistance to the colonial stranglehold on the woods.[18] If more rain was wanted, he

13 *Annales forestières*, vol. 9, 1850, p. 100.

14 Davis, *Resurrecting the Granary of Rome*; Rodolphe Dareste de La Chavanne, *De la propriété en Algérie*, 2nd revised edn, Paris: A. Durand, 1864.

15 Augustin Warnier, *L'Algérie devant l'opinion publique*, Algiers: Molot, 1864, p. 166 and *L'Algérie devant le Sénat*, Paris: Dubuisson, 1863, pp. 46–50. This lobbying paid off: the measures decided ended up being replaced by legislation tailor-made for the colonists. See Davis, *Resurrecting the Granary of Rome*.

16 Cutler, 'Evoking the State Environmental Policy'.

17 Frédéric Robiou de La Tréhonnais, 'L'agriculture en Algérie', *Journal d'agriculture de la ferme*, 5 November 1867, pp. 394–417 (esp. pp. 395–401).

18 Henri Verne, 'La France en Algérie', in *Le Correspondant*, Paris: Charles Douniol, 1869, pp. 1047–8. This social opposition by using fire was also found in the nineteenth century in mainland France, where, as well as featuring in major revolutionary episodes, it was a weapon for the peasantry and agricultural labourers confronting a criminalization of collective usages of nature and competition from mechanization. See Jean-Claude Caron, *Les Feux de la discorde. Conflits et incendies dans la France du XIXe siècle*, Paris: Hachette, 2006.

wrote, it would be necessary not only to repress such practices, but also to implement large-scale reforestation. The deadly catastrophe was treated as one more sign: the 'Arab kingdom' was a kingdom of fire, famine, and death. All power to the colonists!

The latter could count on a forestry authority aligned with their positions. Since its creation in 1846, the Algerian forest administration had gone from strength to strength.[19] Its rangers implemented on the ground the dispossession and criminalization of uses provided for by law. In the 1880s they issued 13,000 violation notices per annum and sometimes resorted to forced labour.[20] Climate decline was one of the major justifications advanced for this policy.

The forestry authority went hand in hand with the League for the Reforestation of Algeria, an association founded in 1881 that attracted some 1,200 members straight away.[21] Its objectives were promoting massive plantations, strengthening forestry laws, and evicting Arabs from wooded regions.[22] On the horizon was the ideal of a climatic restoration. The memory of the great drought of 1866–68 served as a perpetual goad, while scientific legitimacy was provided by authors like Rauch and Surell.

The League was also an echo chamber for a question that had hitherto occupied a back seat in the eco-imperial narrative: the advance of the Sahara. It now seemed vital to counter this menace, for example by planting ramparts of trees on the high plateaux of the South.[23] Some cherished grander dreams: one colonist imagined the Sahara, already dotted with oases, becoming gradually covered with islets of life and humidity, thanks to the digging of artesian wells.[24]

---

19 Lefebvre et al., *Les Eaux et Forêts du XII*ᵉ *au XX*ᵉ *siècle*, pp. 496, 525–6, 558–60.

20 Ibid., p. 560.

21 Anne Bergeret, 'Discours et politiques forestières coloniales en Afrique et à Madagascar', *Revue française d'histoire d'outre-mer*, vol. 80 (298), 1993, pp. 23–47.

22 François Trottier, *Boisement et colonisation*, Algiers: Association ouvrière, 1876, quoted in Bergeret, 'Discours et politiques forestières coloniales', p. 26.

23 Paulin Trolard, *La Colonisation et la question forestière*, Algiers: Imprimerie Casablanca, 1891, p. 80.

24 Ibid.

Threats to the Raj

From the 1840s to the mid-twentieth century, the British Empire was prey to the same climatic anxieties.[25] Declinist narratives justified the dispossession of the natives and the criminalization of their uses of nature. In this case, however, there was no carry-over of worries from Great Britain. The country had not experienced the major climate controversies that shook France in the first half of the nineteenth century. It was the expert opinion and writings produced in the context of the French debates that disseminated the theses of anthropogenic climate action in the Victorian Empire.

This was above all exemplified in India. Although the country possessed immense forests, in the first half of the nineteenth century the British colonizer did not have an assertive forestry policy or a dedicated forest authority. Great Britain's capacities in this area were weak in general: there was no specialist department in the metropolis, either. It was precisely the deficiency of British forestry expertise, compared with the major continental traditions, that amplified the climate alert overseas. The imperial forestry born in Victorian India embraced French climate anxieties, and these helped it impose its authority throughout the Empire.

Alexander Gibson, Edward Balfour, and Hugh Cleghorn were surgeons and naturalists, in charge of the botanical gardens of Bombay, Madras and Mysore respectively. Following a wave of famines in 1838–39, they grew alarmed at the ongoing desiccation of India caused by deforestation.[26] They feared that this fatal process could ultimately jeopardize the very viability of colonization. The connection they established between deforestation and drought was inspired by J.B. Boussingault's article on tropical rains, unquestionably the most influential text for the generalization of climate alarm (Chapter 12).[27]

Initially developed in the context of political struggles over the forests of France, Boussingault's argument now served as a foundation for the diagnoses of imperial foresters. To counter climate decline, they stated,

25 This has been well studied by Richard Grove, *Green Imperialism* and Gregory Barton, *Empire Forestry and the Origins of Environmentalism* (2002), Cambridge: Cambridge University Press, 2004.

26 Grove, 'Conserving Eden', p. 342.

27 Jean-Baptiste Boussingault, 'Memoir Concerning the Effect', pp. 85–106.

the colonial authorities must invest in forest management.[28] Gibson was the first to achieve his ends: in 1847, he secured the creation of a Bombay Forest Department, the first iteration of such an agency in India, and placed himself at its head. To speed things up, from India he pressured the London-based administrators of the East India Company.[29] He warned them that, through its climatic effects, the deforestation of the western coastline posed a grave threat to the company's plans to grow cotton. His argument hit home.

At the same time, a major phase of seizure and guardianship of Indian forests by the British began. In 1855, a Forest Charter confiscated many of the country's forests. The same year, a forestry department was created in Madras on the model of Bombay's, with Hugh Cleghorn at its head. A doctrine of management gradually took shape within the two departments and became officialized in the Indian Forest Service (whose mission was nation-wide), founded in 1864. This doctrine had two main objectives: firstly, to intensify the economic exploitation of certain mountains, such as those supplying the precious hardwood, teak; secondly, to ensure the preservation of climates.

The modalities of forests governance (nationalization, repression of user rights) were conceived as tools for countering climate change. The paradigm of human action on the climate took definitive root in the subcontinent. The rise of imperial forestry technostructure further reinforced a tendency connecting India and the whole Victorian empire to several decades of debate and expert opinion in France on climate change. The paradigm circulated and was self-reinforcing. A number of British experts completed courses at Nancy's Forestry School, where they learned about the ecological decline of the Maghreb. And if they opted to study in Germany, the result was much the same. The bible of German forestry, Heinrich von Cotta's *Grundriss der Forstwissenschaft*, opened with the following assertion: 'When forests are few, rain becomes scarce.' It cited many of the French works prompted by the great inquiry of 1821.[30]

The difference with Algeria is that these theories were now applied on a much vaster scale. In the late nineteenth century, the inspector-general

28 Grove, *Green Imperialism*, pp. 428–73; Barton, *Empire Forestry*, pp. 47–8.

29 Grove, *Green Imperialism*, p. 435.

30 Heinrich Cotta, *Grundriss der Forstwissenschaft*, Dresden and Leipzig: Arnold, 1849, p. 2. He refers here to François Maurel, *Influence météorologique des montagnes et des forêts*, Foix: Pomiès, 1840.

of Indian forests, Berthold Ribbentrop, made his calculations: a 30 per cent rate of forest cover was the correct objective.[31] However, there was a degree of bluff behind these abstract figures and this continental development plan. The control of the British over Indian forests was actually shakier (given the immensity of the expanses) than that of the French in the Maghreb, which had itself been challenged. The aspiration to climate conservation or restoration sometimes seemed like a disproportionate ideal, given the magnitude of the task. Even so, British foresters made great efforts to broadcast their crusade. As in Algeria, discourses on the climate were also a way of addressing the metropolis to demand resources. And, as in the Maghreb, the objective was utilitarian. There was little interest in preserving beauty or harmony. Protecting the forests meant establishing colonial environments that were rational, secure (for the colonists), and productive over time.

In the Raj, we also find an anti-Muslim eco-racism that is like a muted reflection of its counterpart in the French Empire. Thus, for Ribbentrop, the Muslim conquest of the Middle East, Central Asia, and a large part of India was the original disaster that had tipped these regions into a protracted process of deforestation and desiccation.[32] At issue here was domination in the Indian peninsula, first by conquerors from Central Asia and then by the Moghul Empire, until its break-up at the start of the nineteenth century. Muslim expansion was held to have brought fatalism and disdain for property with it – 'communistic precepts' that made forests the property of all.[33] The interpretation of Roman sites by the French in Algeria inspired the British in their take on the ruins found in India. The discourse of collapse went mainstream, along with racist clichés blaming a 'Muslim civilization' charged with every imaginable vice.

The Frontier Climate

The echo of French debates was not only heard on the other side of the Mediterranean after 1850. The climate threat spread as far as the United States, thanks to an author today regarded as one of the founders of

31 Barton, *Empire Forestry*, p. 42.

32 Berthold Ribbentrop, *Forestry in British India*, Calcutta: Office of the Superintendent of Government Printing, 1900, pp. 35–40.

33 Ibid., p. 36.

environmentalism: George Perkins Marsh.[34] From approximately 1860 to 1890, his writings crystallized a surge of disquiet about the climate effects of the colonization of the American West.

Marsh was a philologist and politician who represented the United States abroad several times in the 1850s and 60s, in Turkey, Greece and Italy. He travelled widely in the Mediterranean periphery, where, under the influence of French climatic declinism, nature struck him as injured and demeaned by human action. In 1864, in his book *Man and Nature*,[35] he extended this view to the planet as a whole, offering a compendium of humanity's destructive powers. Marsh listed a whole range of harms to nature, from the extinction of species to pollution. The destruction of forests and its effects on climate occupied a central place, fuelled by the French debates of previous decades.[36] Marsh drew on Moreau de Jonnès, Arago and Becquerel, and discussed the works of Boussingault and Humboldt in detail.

The book enjoyed huge success in the United States. It was at a time when, following the Civil War, the front line of agricultural colonization was reaching the great arid plains of the centre of the country. Paradoxically, but in keeping with the climate optimism of the republic's early decades, some saw Marsh as a new prophet of climate improvement.[37] In 1867, Ferdinand V. Hayden, head of the US Geological Survey in the West, suggested planting an immense curtain of trees on the whole eastern flank of the Great Plains – something that would, he was sure, thoroughly improve their climate. The idea was inspired by *Man and Nature*, where *inter alia* he discovered Boussingault's work.[38]

34 On climate in the historiography of the United States, see the highly stimulating panorama by Sam White, 'Climate, History, and Culture in the United States', *Wiley Interdisciplinary Reviews: Climate Change*, vol. 9, no. 6, 2018 (online publication: DOI: 10.1002/wcc.556; accessed 10 January 2020). On Marsh, see esp. David Lowenthal, *George Perkins Marsh, Prophet of Conservation*, Washington: University of Washington Press, 2000 and Marcus Hall, 'The Provincial Nature of George Perkins Marsh', *Environment and History*, vol. 10, no. 2, 2004, pp. 191–204.

35 George Perkins Marsh, *Man and Nature, or Physical Geography as Modified by Human Action*, New York: Charles Scribner, 1864.

36 In turn, the book became known in France thanks to the geographer Élisée Reclus, 'De l'action humaine sur la géographie physique', *Revue des deux mondes*, 15 December 1864, pp. 762–71. See also Reclus, *La Terre: description des phénomènes de la vie du globe*, Paris: Hachette, 1876, vol. 2, pp. 732–4.

37 Henry Nash Smith, 'Rain Follows the Plow: The Notion of Increased Rainfall for the Great Plains, 1844–1880', *Huntington Library Quarterly*, vol. 10, no. 2, 1947, pp. 169–93.

38 Ferdinand V. Hayden, *First Annual Report of the United States Geological Survey of*

Boosters – entrepreneurs who speculated on the lands of the West – wielded these theories as an exhortation to agricultural colonization. The idea of climate improvement using trees was a way of enhancing the economic potential of those spaces. They also promised that deep ploughing, irrigation, and cultivation would turn the Great Plains green and humidify their climate.[39] In the famous formula by Charles Dana Wilber, an entrepreneur from Nebraska, 'rain follows the plow'.[40] In other words, breaking up the hard soil of the Great Plains would change the water cycle there and hence the climate.

But Marsh's theses also bred *disquiet* about a rapid and excessive deforestation of America, at a time when logging was booming with the rise of rail and steam navigation.[41] This had a very real impact on forestry policy: by warning about the climatic effects of deforestation, Franklin B. Hough obtained a meeting with President Grant and persuaded Congress to create the post of special agent in 1876 – a milestone in the federal assumption of responsibility for forests. Hough's view of the problem was considerably influenced by French works on anthropogenic climate change – he even reproduced in his writings long extracts from Becquerel.[42] Three years earlier, the first major piece of national legislation – the Timber Culture Act of 1873 – aimed to combat the climate threat while ensuring the availability of the resource. It provided for grants of land to farmers engaged in planting and maintaining trees.

---

the Territories, Embracing Nebraska, Washington: US Government Printing Office, 1867, pp. 14–16.

39 Laurence Culver, 'Manifest Destiny and Manifest Disaster: Climate Perceptions and Realities in United States Territorial Expansion', in Christof Mauch and Sylvia Mayer, eds, *American Environments: Climate Cultures Catastrophes*, Heidelberg: Universitätsverlag, Winter 2012, pp. 7–30; David M. Emmons, *Garden in the Grasslands: Boomer Literature of the Central Great Plains*, Lincoln: University of Nebraska Press, 1971.

40 Charles Dana Wilber, *The Great Valleys and Prairies of Nebraska and the Northwest*, Omaha: Daily Republican Press, 1881, p. 143.

41 Both aspects coexist in the torrent of articles that appeared on the links between climate and forests, and which largely drew on the writings of Marsh. See Donald J. Pisani, 'Forests and Conservation, 1865–1890', *Journal of American History*, vol. 75, no. 2, 1985, pp. 340–59.

42 An extract itself summarizing some of the French literature: Franklin B. Hough, *Report on Forestry*, Washington: Government Printing Office, 1878, pp. 310–33. See also Joseph Kittredge, *Forest Influences* [1948], New York: Dove Publications, 1973, p. 9; Vasant K. Saberwal, 'Science and the Desiccationist Discourse of the 20th Century', *Environment and History*, vol. 3, 1997, pp. 309–43 (esp. pp. 314–15); Diana K. Davis, *The Arid Lands*, pp. 135–7.

This wave combining optimism and climatic disquiet ebbed in the 1890s and 1900s, as farmers grew more familiar with the ecological realities of the plains of the West and forestry conservation measures were strengthened. The major federal technical departments also played a part in repudiating the theory of climatic influences. The meteorologists of the Weather Bureau, sceptical about the climatic impact of colonization (Chapter 14), attributed this belief to the copious rainfall experienced by Kansas and Nebraska in the first half of the 1880s, which seemed – wrongly – to bear out the maxim that 'rain follows the plow'. The US Forest Service would make the link between deforestation and flooding its hobbyhorse, but (as in France) on the basis of a pedological interpretation that gradually marginalized the climatic aspect.[43]

Marsh eventually faded from the public sphere and collective memory, before returning to the spotlight decades later and being enthroned (with all the distortions presupposed by this honour) as a 'precursor' of ecological thinking.[44]

From the Sahara to the Namib

By contrast, climate declinism flourished in the British and French empires far into the twentieth century. No longer confined to the Maghreb and India, it now concerned the whole of Africa, from the Cape to the Sahel, from Kenya to Madagascar.

Throughout the continent, severe droughts blighted the decades from 1900 to 1930, driving the colonists of the new territories of French West Africa to despair.[45] The threat of a 'regional', even continental, 'desiccation'

43 Saberwal, 'Science and the Desiccationist Discourse', pp. 315–16, 318.

44 David Lowenthal, 'Marsh and Sauer: Reexamining the Rediscovery', *Geographical Review*, no. 103, 2013, pp. 409–14.

45 On the paradigm of climatic desiccation in French West Africa and the Sahel, see Aziz Ballouche and Aude Nuscia Taïbi, 'Le "dessèchement" de l'Afrique sahélienne: un leitmotiv du discours d'expert revisité', *Autrepart*, no. 65, 2013, pp. 47–66; Joanny Guillard *Au service des forêts tropicales. Histoire des services forestiers français d'outre-mer 1896–1960*, Nancy: AgroParisTech, 2014; Aude Nuscia Taïbi, 'Désertification et dégradation. Ré-interrogation des concepts à la lumière d'exemples africains', Geography HDR, Angers University, 2015; Davis, *The Arid Lands*, pp. 118–25; Richard Grove, 'A Historical Review of Institutional and Conservationist Responses to Fears of Artificially Induced Global Climate Change: the Deforestation–Desiccation Discourse in Europe and the Colonial Context, 1500–1940', in Yvon Chatelin and Christophe Bonneuil, eds, *Nature et*

loomed. People took fright at the encroachment of the Sahara, which might engulf the Sahel. These anxieties were fed by representations originating in the Maghreb: the Saharan question served as the interface between the two edges of the desert. Many colonial officials (administrators, scientists, military men, or foresters) had their first posting in Algeria. With them, and with the expeditions that crisscrossed the continent, circulated the tropes of climatic decline.

The botanist Auguste Chevalier, one of the future strong men of French science, mentioned it in his account of the Chari-Lake Chad expedition (1902–04).[46] The lake, he prophesied, would perish with the extension of the 'Saharan climate': it was doomed to 'complete sterility', on the model of the Baguirmi marshes, which once were lagoons.[47] The report of the major forestry mission of French West Africa carried out some years later was even more alarmist: a veritable requisitory against the practices of the local population (clearing, grazing, bushfires). 'The Sahara', it announced, 'is encroaching southwards; and this as a result of men.'[48] To counter it, it would be necessary, as in Algeria, to prohibit access to deforested zones, restrict pastoral flows, harvest timber methodically, and no longer leave this resource to the 'whims of the blacks'.[49]

In 1913–14, a murderous drought struck the regions of the Sahel. Contemporary climatology regards it as the most severe of the century, exceeding that of the 1970s. It provoked rampant famine.[50] As anthropologists have shown, this disaster left profound traces in the popular memory.[51] Today, the event is still called *diaba* (the great drought) in Bambara, *rafo-manga* (the great famine) or *kitanga* (the great loss) in Fula. A French administrator posted in Bandiagara (Mali) estimated that one-third of the population died of hunger.

Prompted by the catastrophe, a vast inquiry into the desiccation of West Africa was launched in 1917. At the helm was Henry Hubert, an

---

environnement, vol. 3, *Les Sciences hors d'Occident au XXe siècle*, Paris: ORSTOM, 1996, pp. 155–74.

46 August Chevalier, *Mission Chari-lac Tchad, 1902–1904. L'Afrique centrale française*: récit du voyage de la mission, Paris: A. Challamel, 1907.

47 Ibid., pp. 336–7.

48 Jean Vuillet and P. Giraud, 'Mission forestière de l'Afrique Occidentale: Rapport d'ensemble', *L'Agriculture pratique des pays chauds*, 1909/2, pp. 58–74 (text in six parts).

49 Ibid., p. 66.

50 Guillard, *Au service des forêts tropicales*, p. 143.

51 Ballouche and Taïbi, 'Le "dessèchement" de l'Afrique sahélienne'.

alumnus of the colonial school and a geologist, trained in the Muséum and then in the field in Africa (he would become patron of the 'colonial meteorological department' created in 1829). According to him, the desiccation of West Africa was a process that had begun a very long time ago but had accelerated in recent decades. Everything pointed to it, from the reduction in the flow of rivers to the fall in groundnut yields, despite efforts to intensify production for export.[52] The testimonies he collected in French West Africa tended in the same direction; they also showed the prevalence of climate anxiety among colonial elites, always dreading new famines like that of 1913–14.[53]

Among colonial foresters there was no doubt either: deforestation was underway everywhere, from West Africa to Madagascar, which explained the reduction in rains, the river spates, and the erosion they observed.[54] These arguments were used on the ground by forestry managers in their constant conflict with European firms – a tricky struggle in which they had to counter powerful economic interests, with strong links to the authorities, that took a dim view of any extension (however modest) of regulations.[55]

The spectre of climate change also haunted the British colonies in Africa. As in the French Empire, anxiety soared during the 1920s and 30s. Of especial concern was the extension of the Kalahari Desert – one million square kilometres shared between Namibia, South Africa and Botswana. The natives' subsistence practices came under the spotlight here as well. In the 1920s, a South African Drought Commission was founded to assess the situation,[56] and the South African geologist Ernest Schwarz sought to

52 Henry Hubert, 'Progression du dessèchement dans les régions sénégalaises', *Annales de géographie*, no. 143, pp. 376–85 and 'Le dessèchement progressif en Afrique-Occidentale', *Bulletin du Comité d'études historiques et scientifiques de l'Afrique-Occidentale française*, 1920, pp. 401–67.

53 See, for example, Hubert, 'Le dessèchement progressif en Afrique-Occidentale', p. 444.

54 For a good summary of this state of mind, see Maurice Mangin, 'Une mission forestière en Afrique-Occidentale française', *La Géographie*, vol. 42, no. 4, 1924, pp. 449–83, 629–54. On Madagascar, see Davis, *The Arid Lands*, pp. 100, 124. For further examples, see Guillard, *Au service des forêts tropicales*.

55 Yannick Mahrane, Frédéric Thomas and Christophe Bonneuil, 'Mettre en valeur, préserver ou conserver? Genèse et déclin du préservationnisme dans l'empire colonial français (1870–1960)', in Charles-François Mathis and Jean-François Mouhot, eds, *Une protection de la nature et de l'environnement à la française? (XIXe–XXe siècle)*, Seyssel: Champ Vallon, 2013, pp. 62–80.

56 Richard Grove and Vinita Damodaran, 'Imperialism, Intellectual Networks, and

raise public awareness about what he called the 'Kalahari Problem'.[57] To solve this he proposed a 'Kalahari Scheme' of his own devising, which consisted in reforesting the land and diverting the outlets of southern Africa's major rivers, from the sea to the desert – a continental geo-engineering that seems like a grotesque inflation of the concrete policies of climate-oriented forest conservation.

Further north, the forestry departments recently created in British East Africa (Kenya and Uganda), in Nigeria and in Sierra Leone were fraught with the same anxieties. In the Sudan under British hegemony, as in French West Africa, there were fears of being engulfed by the Sahara. All the British colonies in Africa shared the same fear of anthropogenic climate change, and all accused the Africans themselves of causing it. A little earlier in the century, the German colony of Namibia (1884–1915) reverberated with the same warnings.[58] The scenario of plant and climate degradation at the hands of local populations, a fixed idea since the early years of colonization, motivated a host of ambitious proposals for tree-planting. One of them, citing the successes of British foresters in the Cape region, envisaged creating a green belt of forests along the coast, to master the Namib desert within 100 years.

At the start of the twentieth century, torn between anxieties and unbridled hubris, the whole world of European colonization considered itself at war against climate decline.

A Planet of Deserts

These climate anxieties peaked in the second half of the 1930s – a period that also saw the first signs of their waning. They peaked because the encroachment of the Kalahari, the Sahara and the Indian deserts, added to the droughts in French West Africa, Sudan, and Madagascar, were mobilizing colonial elites, experts and European states; and because representations, scientific discourse and managerial practices were circulating from colony to colony, within and between empires.

Environmental Change: Origins and Evolution of Global Environmental History, 1676–2000', *Economic and Political Weekly*, no. 41, 2006, pp. 4345–54 (p. 4348).

57 Ernest H.L. Schwarz, *The Kalahari or Thirstland Redemption*, Cape Town/Oxford: Maskew Miller/Blackwell, 1920, p. vi.

58 Harri Olavi Siiskonen, 'The Concept of Climate Improvement: Colonialism and Environment in German South West Africa', *Environment and History*, vol. 21, no. 2, 2015, pp. 281–302.

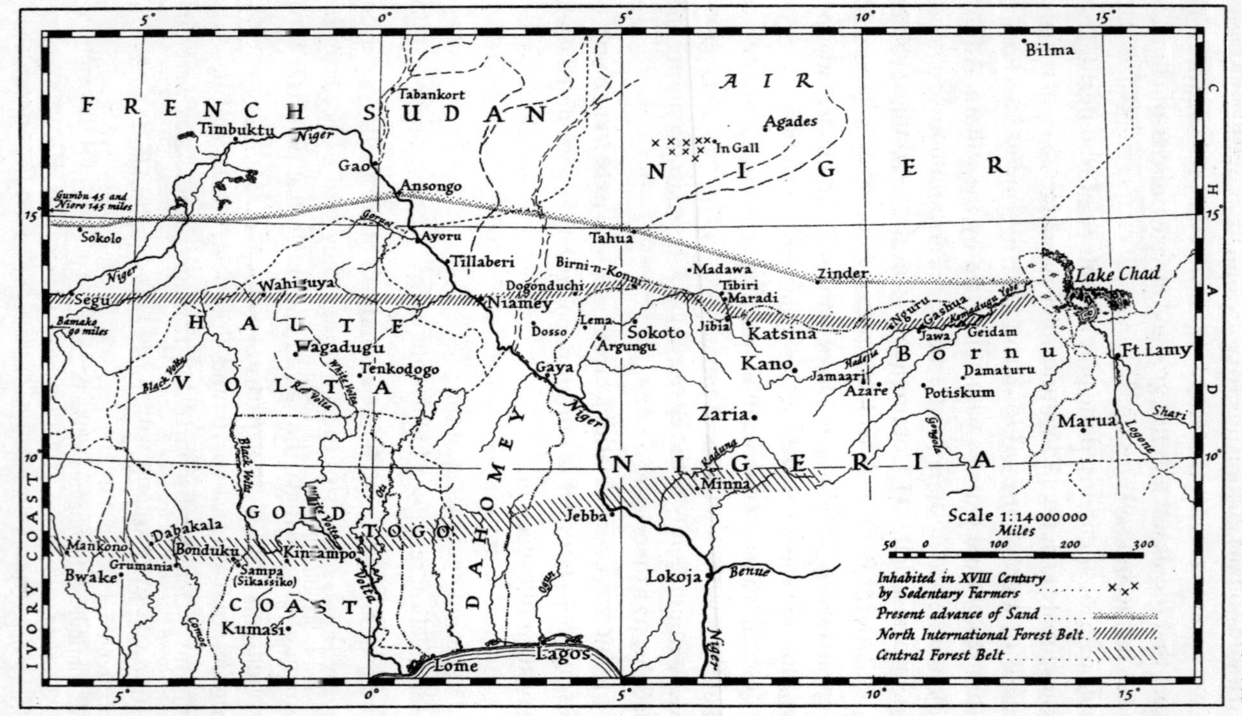

The encroachment of the Sahara and the plan for two belts of forests to stop it and improve the climate of the British and French colonies of West Africa. Edward P. Stebbing, 'The Encroaching Sahara: The Threat to the West African Colonies', *Geographical Journal*, vol. 85, no. 6, 1935, pp. 502–24 (p. 508).

At the same time, the thesis of climate change was being increasingly disputed. When, in 1935, the prominent British forester Edward Stebbing published a study on the encroachment of the Sahara caused by deforestation, his work was vigorously criticized.[59] An official report contradicted it, asserting that the desert was not growing and that the real issue was not climate change, but soil erosion.[60] Stebbing of course stood firm, but a multitude of critics denounced the prophets of doom who piped up as soon as the climate experienced naturally dry sequences.[61]

The thesis of a deterioration in African environments was deeply rooted among colonial experts who, after countries gained independence, became involved in the programmes of the Commonwealth and the Union française, later the Communauté française, working in 'development' activities.[62] But a link with anthropogenic *climate* change was increasingly in doubt. Thus, on pondering the issue in 1950, the naturalist Théodore Monod (a great walker of the Sahara and the top French specialist in deserts) was sceptical. Citing South African and British studies, he stressed their conclusions regarding the lack of impact of forests on rains, the extreme antiquity of the Egyptian desert, and the outlandishness of Ernest Schwarz's theories about the Kalahari.[63]

At this point the United Nations began to investigate the subject of arid zones and deserts, under the impetus of the USA and experts trained in the colonial empires. In 1951, UNESCO launched an international research programme around this theme. Its work unfolded over fifteen years and explored every facet of the problem. The notion of anthropogenic desertification remained highly influential, but climate change was no longer the key issue.

59 Edward P. Stebbing, 'The Encroaching Sahara: The Threat to the West African Colonies', *Geographical Journal*, vol. 85, no. 6, 1935, pp. 506–24.

60 F.S. Collier and J. Dundas, 'The Arid Regions of Northern Nigeria and the French Niger Colony', *Empire Forestry Journal*, February 1937, pp. 184–93.

61 Edward P. Stebbing, 'The Man-Made Desert in Africa: Erosion and Drought', *Journal of the Royal African Society*, vol. 37, no. 146, 1938, supplement, pp. 3–40 (esp. pp. 12–14, 20); Brynmor Jones, 'Desiccation and the West African Colonies', *Geographical Journal*, vol. 91, no. 5, 1938, pp. 401–23 (esp. p. 417).

62 This continuity is highlighted by Davis, *The Arid Lands*.

63 Théodore Monod, 'Autour du problème du dessèchement africain', *Bulletin de l'IFAN*, April 1950, pp. 514–23. Another influential expert, André Aubréville, believed in the climatic effects of deforestation, but lost himself in circumlocutions explaining that they could not be proven empirically: *Climats, forêts et désertification de l'Afrique tropicale*, Paris: Éditions géographiques, maritimes et coloniales, 1949, pp. 329–44.

How are we to explain this waning? The old empires had disintegrated, and with them, their climatic paradigm. The new global power – the American empire – had its own ecological narratives, its own theories and its own practices, rooted not in the experience of deforestation but in the trauma of the Dust Bowl – the catastrophic combination of erosion and dust storms that ravaged the Great Plains in the 1920s and 30s.[64] The government had then instituted the powerful Soil Conservation Service, whose methods of soil protection and restoration became the benchmark after 1945 – not only in the UN framework, where the US was dominant, but also in some Old World colonies. In the Maghreb, the Eaux et forêts authority embarked on robust programmes of 'soil defence and restoration' in the 1940s and 50s, using techniques of levelling and afforestation borrowed from the Soil Service.[65]

In disseminating a concept of conservation focused as of the 1930s on the importance – physical, agricultural, even cultural – of soils, American influence diminished the idea of human climate action. One collapsology drove out the other. In 1935, the head of the Soil Service, Hugh H. Bennett, explained that 'the world is strewn with the ruins of once flourishing civilizations, destroyed by erosion', and that this threatened America too.[66] His deputy, Walter C. Lowdermilk, was known around the world for his articles on erosion and the collapse of societies such as the Maya, or the ancient kingdom of the Queen of Sheba.[67] Among the photographs Lowdermilk projected during lectures, one shows depleted soils in Alabama; he urged immediate action to halt those processes in America. Another was of the Roman ruins of Timgad in Algeria. If action were not taken, this is what could happen to American society: as in the Maghreb, it could self-destruct in the wake of veritable ecological suicide.[68]

Lowdermilk was perfectly clear: climate change played a marginal, not to say non-existent, role in the processes of desertification. The desiccation

---

64 Donald Worster, *Dust Bowl: The Southern Plains in the 1930s*, New York: Oxford University Press, 1979.

65 Davis, *The Arid Lands*, pp. 146–7.

66 Quoted by Frank Uekötter, 'The Meaning of Moving Sand: Towards a Dust Bowl Mythology', *Global Environment*, vol. 8, no. 2, 2015, pp. 349–79.

67 Walter C. Lowdermilk, 'Man-Made Deserts', *Pacific Affairs*, vol. 8, no. 4, 1935, pp. 409–19. On Lowdermilk, see Douglas Helms, 'Walter Lowdermilk's Journey: Forester to Land Conservationist', *Environmental Review*, vol. 8, no. 2, 1984, pp. 132–45.

68 Walter C. Lowdermilk, 'AGU Evening General Assembly. Presidential Address. Down to Earth', 2 June 1944.

The ruins of the old Roman town of Timgad (Algeria). Lowdermilk comments on this image in his writings, highlighting the erosion of the soils nearby, and their current inability to support more than a few hundred inhabitants. Walter C. Lowdermilk, *Conquest of the Land through 7000 Years* [1948], Washington, DC: US Dept. of Agriculture – Soil Conservation Service, 1953, p. 17.

of soils due to erosion might sometimes give the impression that the climate had changed, but this was not the case.[69] The impact of the condition of the terrain on the *circulation* of water was sufficient to explain the observations. He acquired this conviction during a long stay in China in the mid-1920s. There, in the eroded landscapes of the Shanxi, he noticed that some zones had remained fertile and wooded: the compounds of temples, where vegetation had been preserved. This proved to him the climate had not changed: it was the soil that underwent the effects of deforestation, and then had an impact on waterways.[70] The thin layer of earth was everything in human life. The fate of civilizations was determined, far more than anyone had imagined, by what happened to raindrops once they hit the ground.[71] The fate of a drop of water made and unmade empires.

This pedological paradigm would play an important role in the rise of environmentalism in the United States. In 1948, two books appeared that became bestsellers and were translated into several languages: *Road to Survival*, by William Vogt, and *Our Plundered Planet*, by Fairfield

69 Lowdermilk, 'Man-Made Deserts', p. 410.
70 Lowdermilk, 'AGU Evening General Assembly', p. 207.
71 Ibid., pp. 208–9.

Osborn.[72] As well as contributing to the success of neo-Malthusianism in the States, both focused on the issue of soils and their destruction on a planetary scale. They also fuelled the new pedological interpretation of global ecological damage. Osborn's book even devoted a chapter to 'the new geological force: man' (the phrase was from the Russian scientist, Vladimir Vernadsky). We still find incidental references to possible climate changes,[73] but the main problem lay elsewhere: in the nurturing substratum that is the soil, which had to be protected if humanity was to survive. Climate, formerly the marrow of Buffon, Rauch and Ribbentrop's happy or tragic anthropocene, was no longer central to that of Osborn and Vogt.

72 William Vogt, *Road to Survival*, New York: W. Sloane, 1948; Fairfield Osborn, *Our Plundered Planet*, Boston: Little, Brown and Co., 1948. See Fabien Locher, 'Fairfield Osborn', in Dominique Bourg and Alain Papaux, eds, *Dictionnaire de la pensée écologique*, Paris: PUF, 'Quadrige', 2015.

73 Vogt cites the hypothesis of a shift in African climates, but as a subsidiary argument, in *Road to Survival*, pp. 245–8.

16

The Innocent Carbon of the Nineteenth Century

In 1832, the mathematician and inventor Charles Babbage pondered the development of the copious quantities of carbon emitted into the atmosphere by British industry. Steam engines, he correctly noted, 'are constantly increasing the atmosphere by large quantities of carbonic acid and other gases noxious to animal life. The means by which nature decomposes these elements, or reconverts them into solid form, are not sufficiently known'.[1] In France Eugène Huzar, an alarmist philosopher largely forgotten today, made the following prediction in 1857:

> In one hundred or two hundred years, being crisscrossed with railways and steam ships and covered with mills and factories, the world will release billions of cubic metres of carbon dioxide and carbon monoxide. And since the forests will have been destroyed, these hundreds of billions of cubic metres of carbon dioxide and carbon monoxide may well disturb the harmony of the world somewhat.[2]

Premonitory as they may seem, these warnings have nothing to do with our contemporary climate disaster. At the time when Babbage or Huzar

1 Charles Babbage, *On the Economy of Machinery and Manufactures*, London: Knight, 1832, p. 17.

2 Huzar, *L'Arbre de la science*, p. 106. For Huzar, the increase in CO_2 was a hygiene problem.

were expressing concern at the atmospheric impact of carbon, no one yet imagined that CO_2 could affect global temperatures. More surprisingly, such anxieties were not particularly original: a whole host of scientists had studied the carbon cycle throughout the nineteenth century and, in subsidiary fashion, the impact of industry on it. Some were preoccupied with the fertility of land and with agricultural chemistry, and tracked the constitutive chemical elements in harvests; others looked into the flora of the Carboniferous or the formation of sedimentary rocks; yet others tried to explain the stability of the atmosphere or to prove the existence of God. At all events, however, climate change was not their problem. Anthropogenic disruption of the water cycle was a source of unease in the nineteenth century; this was much less true of the carbon cycle. And, even after John Tyndall and Svante Arrhenius had highlighted the role of CO_2 in the greenhouse effect, the latter would be deemed positive for the climate and for humanity until the middle of the twentieth century.

Carbon Theology

In 1787, when the word 'carbon' was coined,[3] the idea of an offset between the respiration of animals generating carbon dioxide and that of plants absorbing this acid was widespread among chemists.[4] Lavoisier was already amazed at the 'marvellous circulation between the three realms'.[5] A chemical vision of the world became established that pictured life as a set of material circulations. And, throughout the nineteenth century, this vision stirred the imagination of natural theologians: the divine order could be read more clearly at the level of the constituents of matter and in the language of the new nomenclature. Just as Newtonian astronomy detected God's hand in the stability of orbits, so post-Lavoisian chemistry claimed to read the divine design in the constant circulation of chemical elements. Nature was an infinitely complex clock, resting on

3 Louis-Bernard-Guiton de Morvaux, Antoine-Laurent Lavoisier, Claude-Louis Berthollet and Antoine de Fourcroy, *Méthode de nomenclature chimique*, Paris: Cuchet: 1787, p. 117.

4 On the history of the elucidation of the carbon cycle, see Matthieu Emmanuel Galvez and Jérôme Gaillardet, 'Historical Constraints on the Origins of the Carbon Cycle Concept', *C. R. Geoscience*, no. 344, 2012, pp. 549–67.

5 Quoted in Jean-Paul Deléage, *Histoire de l'écologie*, Paris: La Découverte, 1991, p. 51.

countless flows of forms of matter linking living beings with one another and with the mineral world. These 'cogs', 'wheels' or 'cycles' conduced to the smooth operation of the whole and to the preservation of the world's habitability. At the midpoint of the nineteenth century, works of popularization, sermons and public lectures undertook to prove God's existence by exhibiting the material arrangement of Creation thanks to chemistry.[6]

Take, for example, *The Chemistry of Common Life* by the Scot James Finlay Johnston, one of the most accomplished works of this kind. It successively sets out the cycles of matter essential to life: water, carbon and nitrogen. God had organized them so as to preserve the chemical balance of the world indispensable to life. According to the author, the carbon cycle was the most 'beautiful', the most mysterious, the most subtle, the one whose apologetic significance was clearest. For carbon was to be found in the atmosphere in a proportion at once minuscule and constant – and this when gigantic masses of it were stored in coal, dissolved in the oceans, or trapped in limestone rock.[7] The theological interpretation of this cycle was ambivalent. For if, on the one hand, it seemed to prove the existence within Creation of a 'good purpose', well-intentioned towards Man,[8] on the other, it rendered the very existence of life on Earth fragile and fleeting. According to Johnston, life 'is a little episode, so to speak, in the great poem of creation'.[9] According to another theologian-chemist, 'in all the arrangements of the material world whose details science has brought to light . . . the slightest imaginable change may often be strictly shown to be incompatible with the safety and equilibrium of the whole.'[10]

These worries, formulated in connection with the carbon cycle, remained fairly abstract. In the nineteenth century, the major political

6 The great English theologian Thomas Chalmers explained that it was 'most of all in chemistry where the internal processes of nature are found to possess a beauty'. Cf. Chalmers, *On the Power, Wisdom and Goodness of God as Manifested in the Adaptation*, London: Pickering, vol. 1, 1834, p. 193. Reference to Chalmers is frequent. See, for example, *Religion and Chemistry, Or Proofs of God's Plan in the Atmosphere*, New York: Scribner, 1864.

7 James Finlay Johnston, *The Chemistry of Common Life* [New York: Appleton, 1871, vol. 2, p. 350ff. According to Johnston, the residues of past marine life represented two-fifths of the carbon on the planet.

8 Ibid., p. 363.

9 Ibid., p. 366.

10 George Fownes, *Chemistry, as Exemplifying the Wisdom and Beneficence of God*, New York: Moore, 1844, p. 34.

question about the chemical balance of the world – a question refracted in problems of enormous scope – concerned neither the atmosphere nor carbon. It involved the soil and the cycles of nitrogen, potassium and phosphorous. Urbanization crystallized the anxieties of chemists (including the famous Justus von Liebig) or socialist thinkers (including Karl Marx): the concentration of human beings and their excreta prevented the mineral substances indispensable to their fertility being restored to the fields. The disposal of human faeces was at the heart of key political debates: over the social question (it was thought that the impoverished soil of the countryside fomented hunger, pauperism and revolution); over collapse (according to Liebig, Rome fell for failing to manage its excreta); over geopolitics, on account of the monopolization of Peruvian guano by the British; and finally over urban hygiene, and by extension the degeneration of the people.[11]

The alteration of the carbon cycle did not induce the same disquiet, or, when unease was expressed, it was not so much about human emissions as about the atmosphere's *natural* tendency to carbon depletion. According to Johnston, left to itself vegetation would absorb the totality of atmospheric carbon in just twenty-two years. According to Robert Ellis, author of *The Chemistry of Creation*, this fatal purge might occur in a mere ten. Fortunately, animal respiration, volcanism, and human industry – 'an important agent in promoting the circulation of carbon on the globe'[12] – returned the carbon trapped by vegetation to the atmosphere.

At the height of industrialization, this theology of carbon naturally had a justificatory role: in the face of obvious industrial pollution and mass complaints, chemists could argue for the quasi-providential role of factories. Let us take an example. From 1852 onwards, revolt was rumbling against the chemical factories in the Charleroi region of Belgium: potato harvests promised to be disastrous and the peasants blamed, not blight, but the acid fumes given off by soda factories. They called on the authorities to close them down. During a demonstration, troops opened fire and there were two deaths. Léon Peeters, a pharmacist from

11 John Bellamy Foster, *Marx's Ecology: Materialism and Nature*, New York: Monthly Review Press, 2000; E. Marald, 'Everything Circulates: Agricultural Chemistry and Recycling Theories in the Second Half of the Nineteenth Century', *Environment and History*, vol. 8, no. 1, 2002, pp. 65–84.

12 Johnston, *Chemistry of Common Life*, p. 631. According to Robert Ellis, even the consumption of tobacco played a role in maintaining the carbon cycle: see his *Chemistry of Creation: Being an Outline of the Chemistries of the Earth*, London, 1852, p. 364.

Charleroi, was imprisoned for publishing a pamphlet accusing the factories of destroying the harvests. This was the tense context in which a chemistry professor at Brussels University, Corneille-Jean Koene, gave a series of public lectures. Their title was eloquent: *On Creation, from the formation of the Earth to the extinction of the human species, or an overview of the natural history of air and noxious air in connection with acid factories and the complaints about their works.*[13] He told his doubtless bemused listeners that chemical factories, far from destroying harvests, helped regulate the overall composition of the atmosphere. His reasoning was as follows: the increase in the human population, its livestock and its buildings, sequestered carbon and raised the atmosphere's oxygen content, rendering the population more 'sensual, sensitive and capricious'. By burning coal, industry stabilized the atmosphere's carbon content and arrested the fermentation of minds and the feminization of bodies. Even better: the hydrochloric acid given off by the much-maligned soda factories destroyed alkaline miasmas and thus reduced the risk of epidemics. In short, factories contributed to human health and the correct balance of Creation alike.

Regulatory Mechanisms

The homogeneity and stability of the atmosphere was an important issue for nineteenth-century chemistry. The most illustrious experimenters – Cavendish, Davy, Thénard, Gay-Lussac, Humboldt, Boussingault and Regnault, among others – expended considerable efforts in analysing atmospheric air with the greatest possible precision, for the four corners of the Earth and for different epochs. Their goal was to ascertain that the atmosphere's composition was everywhere the same, and that it remained stable. The idea was also to furnish a reference-point for detecting future developments. Humboldt and Gay-Lussac explained that their analyses were addressed to future scientists as much as to their contemporaries: 'It would be useful for future centuries to clearly record the physical condition of the globe today . . . to fix authentically . . . the average temperature of each climate *and the proportion of the main constituents of the atmosphere*' (our emphasis).[14]

13 Brussels: Larcier, 1856.

14 Alexander Humboldt and Joseph Louis Gay-Lussac, 'Expériences sur les moyens

Thirty-five years later, Jean-Baptiste Dumas and Boussingault, recapitulating the numerous analyses of the atmosphere, concluded that scientists working in different times and places had hit upon identical results: counterintuitive at the time, the idea of the atmosphere as a global, stable and perfectly homogenous object became ingrained.[15] All of them – Dumas, a fervent Catholic, in particular – marvelled at the immutability of the atmosphere 'in the immensity of centuries'.[16] According to them, this stability was attributable to a mass effect: 'The phenomena of organic life . . . the combustions that occur on the Earth's surface, all those events that our imagination likes to magnify . . . pass virtually unperceived as regards the general composition of the air surrounding us.'[17] Human activity, and even vegetation, did not alter the chemical proportions of the global atmosphere, so voluminous was it. In 1855, the chemist Eugène Péligot took up this idea. Having calculated that European industry annually injected 80 billion cubic metres of carbon dioxide into the atmosphere, or the equivalent of the respiration of 500 million individuals, he reassured his readers that 'these quantities, considerable as they appear to us, are surely nothing compared with the immensity of our atmosphere'.[18] The sky was vast enough for the gaseous residues of industry to be discharged into it, risk-free.

The more subtle idea of a stability generated not by mass, but by self-regulating planetary processes, emerged from geological questions. The link between the history of the atmosphere and that of the Earth was part of the common baggage of scientists. For example, the fact that limestone rocks contained 'still air' – the future CO_2 – had long been known: a classic way of producing that gas was to make sulfuric acid react on limestone. Chemists also knew that volcanos ejected enormous quantities of CO_2.[19] How, then, to account for atmospheric stability over a long period, despite

eudiométriques et sur la proportion des principaux constituants de l'atmosphère', *Journal de physique*, vol. 60, 1805, pp. 129–67.

15 On this important paper, see Alexis Zimmer, *Brouillards toxiques, vallée de la Meuse, 1930, contre-enquête*, Brussels: Zones sensibles, 2017, pp. 150–4.

16 Jean-Baptiste Boussingault and Jean-Baptiste Dumas, *Essai de statique chimique des êtres organisés*, Paris: Fortim, 1844, p. 7.

17 Boussingault and Dumas, 'Recherche sur la véritable constitution de l'air atmosphérique', *Comptes rendus des séances de l'Académie des sciences*, vol. 12, 1841, p. 1021.

18 *L'Ami des sciences*, vol. 1, 1855, p. 174.

19 Boussingault analysed the 'elastic fluids' given off by volcanos during his stay in Bolivia. They were 95 per cent composed of carbonic acid. Cf. *Annales de chimie et de physique*, vol. 62, 1833, p. 8. According to Boussingault and Dumas, this volcanic activity

disturbances – volcanic eruptions, carbon sequestration by plants – as massive as they were innumerable?

The answer to this question lay in the discovery of the geological cycle of carbon, and emerged from a seemingly unrelated activity: the porcelain industry. Jacques-Joseph Ebelmen was a mining engineer.[20] In 1847 he was director of the national porcelain factory based in Sèvres. The raw material of porcelain is a very particular white clay: kaolin.[21] Seeking to elucidate the origin and formation of kaolin, Ebelmen discovered that the degradation of primary rocks into clay stored gigantic quantities of carbonic acid, equivalent to more than 100,000 times that contained in the atmosphere. The stability of the atmosphere's make-up, then, did not depend on the organic carbon cycle, but was linked instead to geological processes. If a volcano inundated the atmosphere with CO_2, vegetation would certainly be stimulated, Ebelmen said, but the regulatory mechanism would be essentially maintained through the faster decomposition of the rocks caused by carbonic acid.[22]

It is to another French chemist, Théophile Schloesing, that we owe the identification of the second major factor regulating the atmosphere – namely, the oceans. A graduate of the École Polytechnique, professor at the National Institute of Agronomy, Schloesing, like numerous chemists, was interested in the cycles of matter, nitrogen in particular. In 1880, he demonstrated experimentally that the constant proportion of CO_2 in the atmosphere was created by the interplay of the formation and dissociation of bicarbonate in water. The ocean, he wrote, was 'a powerful regulator' of the quantity of atmospheric CO_2.[23] The article made a big splash. It was summarized in several medical publications and was even mentioned in the *Revue des Deux Mondes*. The analogy between the regulation of the Earth and that of bodies was striking. Albert Dastre, professor of

could have played a role in very remote epochs, but in the course of the Earth's history the compensatory activity of plants resulted in a balanced atmosphere.

20 His seminal article of 1845 was rediscovered a little while ago by the geochemists Robert Berner and Jérôme Gaillardet. We are grateful to the latter for this valuable clarification. Jacques-Joseph Ebelmen, 'Recherches sur les produits de la décomposition des espèces minérales de la famille des silicates', *Annales des Mines*, vol. 7, 1845, pp. 3–66.

21 Kaolin is a strategic material studied by numerous chemists – for example, the mineralogy professor at Lyon Joseph Fournet, the chemist Davy in England, or Alexandre Brongniart and Faustino Malaguti at the Sèvres factory.

22 Ebelmen, 'Recherches sur les produits de la décomposition', p. 65.

23 Théophile Schloesing, 'Note sur la constance de la proportion d'acide carbonique dans l'air', *Comptes rendus de l'Académie des sciences*, vol. 90, 1882, p. 1411.

physiology at the Sorbonne, stressed the importance of Schloesing's discovery: the stability of the atmosphere rested not only on the organic carbon cycle – a precariously balanced one – but on the whole surface of the ocean. Dastre wrote that 'we needed some kind of regulatory mechanism acting automatically, which might establish this strict correspondence between profit and loss. We suspected its existence and Mr Schloesing has revealed it to us.'[24] In the last third of the nineteenth century the idea took root therefore that the atmosphere was stable thanks to life, rocks, and oceans. Its development was entwined with that of the Earth and the organic world. According to Ebelmen, 'the variations in the nature of the air have without doubt constantly been related to the organized beings that lived in each epoch.'[25]

Once again, however, no one was linking CO_2 to the global temperature. Ebelmen is probably the person who, prior to the US scientist Eunice Foote and the physicist John Tyndall, went furthest in this direction, invoking not the greenhouse effect but the increased density of the atmosphere. Echoing the theses of his friend Adolphe Brongniart – namely, that it must have been at least 8° hotter in the Carboniferous than today (see Chapter 14) – Ebelmen explained that the great quantity of atmospheric CO_2 was bound to result in 'a heavier gaseous envelope' and 'correspond to a higher concentration of solar heat'.[26]

Precursors of Their Time

This book has described the political, theological, imperial, and scientific contexts within which climate change has been perceived, thought, anticipated, feared, and endured, but also celebrated, since the sixteenth century. And yet a different narrative is possible, one detaching from the maelstrom of nineteenth-century theories – about heat, the physics of the globe, ice ages, or gas spectrometry – the scientific advances that have led to the contemporary diagnosis of global warming. This history is infinitely clearer and shorter, but also more reassuring. The relevant

24 'L'atmosphère, sa constitution, les nouveaux gaz', *Revue des Deux Mondes*, vol. 148, 1898, pp. 198–215.

25 Ebelmen, 'Recherches sur les produits de la décomposition', p. 66.

26 Ibid., p. 65. Ebelmen probably derived this idea from Robert Harkness, 'The Climate of the Coal-Epoch', Manchester, 1843, translated the following year in the *Bibliothèque universelle de Genève*, vol. 49, 1844, pp. 358–64.

knowledge has been validated by contemporary physics; doubt has been banished from it, along with politics, conquests, forests, bad harvests, the people, hunger, and revolts.

This second history has been written 'live' by climatologists themselves. As early as 1906, the Swede Svante Arrhenius established the official genealogy of the greenhouse effect: Jean-Baptiste Fourier, John Tyndall, and himself. It was taken up by the pioneers of global warming – the British engineer Guy Steward Callendar in 1938, the Canadian physicist Gilbert Plass in 1956,[27] until from the 1970s, and in a different context, it was reprised by virtually all climatologists.[28] Obviously, this history aimed to create scientific authority. Claiming to walk in the glorious footsteps of Fourier, Tyndall and Arrhenius (the first two count among the major physicists of the nineteenth century, and Arrhenius won the 1903 Nobel Prize for chemistry) reinforced the credibility of the thesis of anthropogenic change, at a time when it was not yet widely accepted. This use of the figure of the precursor is perfectly commonplace in science. But it is a poor starting place for anyone wishing to understand the true aims of Fourier, Tyndall and Arrhenius.

Let us begin with Fourier. His 1824 article 'sur les températures du globe terrestre et des espaces planétaires' is constantly cited by historians and climatologists as the starting-gun of the theory of the greenhouse effect, and by extension, of the diagnosis of global climate change.[29] It is true that, in this article, he describes the warming that could be caused by the difference in transmission, by atmospheric layers, on the one hand of solar rays and on the other of the heat beamed back from the Earth by radiation.[30] To give tangible form to the process, he compared the planet and

27 Gilbert N. Plass, 'Effects of Carbon Dioxide Variation on Climate', *American Scientist*, vol. 44, no. 3, 1956, pp. 302–16.

28 H.E. Landsberg, 'Man-Made Climatic Changes: Man's Activities Have Altered the Climate of Urbanized Areas and May Affect Global Climate in the Future', *Science*, vol. 170, 1970, pp. 1265–74; B.J. Palmer, 'A Review Paper on the Effect of Carbon Dioxide and Aerosols on Climate Modification', *Environmental Letters*, vol. 5, no. 4, 1973, pp. 249–65.

29 James R. Fleming, *Historical Perspectives on Climate*, pp. 55–64. On Fourier as precursor, see Spencer R. Weart, *The Discovery of Global Warming*, pp. 2–3; Jacques Grinebald, 'De Carnot à Gaïa, l'histoire de l'effet de serre', *La Recherche*, no. 243, 1992, pp. 532–8; *Sur les origines de l'effet de serre et du changement climatique*, Paris: Éditions La ville brûle, 2010, pp. 9–50.

30 Jean-Baptiste Joseph Fourier, 'Remarques générales sur les températures du globe terrestre et des espaces planétaires', *Annales de chimie et de physique*, vol. 27, 1824, pp. 136–67. A subsequent modified version is equally cited: 'Mémoire sur les températures

its atmosphere to a vase, enclosed by a series of glass panels and exposed to the sun. But he neither mentioned the human activity that might increase the temperature nor used the term 'greenhouse effect', which arose much later, first appearing in English in the 1900s.[31]

This research into the thermal balances of the globe did not proceed from abstract intellectual curiosity. Nor was Fourier a precursor of climate change in the contemporary sense. If he intervened in 1820, and again in 1824, it was to respond to the concerns of the society *of his time* about a possible deterioration in European climates. His works sought to allay the anxieties aroused by the sequence of the Tambora aftermath and the heavy winter of 1819–20: the threat of Buffon's cooling, the vertigo induced by the possible climatic effects of deforestation. Fourier was well placed to grasp the scale of these alerts. Between 1802 and 1815, at the time when he produced his great theory of heat, he was prefect of the Isère, dealing day in, day out with the problems posed by clearances, flooding and mountain erosion. Then he was secretary of the Académie des sciences when it was tasked by the Interior Ministry with collating the responses to the inquiry of April 1821.

If Fourier refuted Buffon's theses, it was to point more firmly to different causes:

> *The establishment and progress of human societies*, the action of natural forces, can significantly, and over vast areas, change the condition of the surface of the ground, the distribution of water and the great movements of the air. Such effects are apt, over the course of several centuries, to induce variations in the degree of average warmth.[32]

His equations contained a coefficient that expressed the *condition of the Earth's surface*, a coefficient whose value impacted on the temperature. For Fourier, this confirmed mankind's possible active impact on the climate. For, he explained, 'the condition of the surface can undergo accidental

du globe terrestre et des espaces planétaires', *Mémoires de l'Académie des sciences*, vol. 7, 1827, pp. 570–604.

31 We have identified its first occurrence in the physicist John H. Poynting: 'The "*Blanketing effect*", or as I prefer to call it, the "green-house-effect" of the atmosphere.' Poynting, 'On Prof. Lowell's Method for Evaluating the Surface-Temperatures of the Planets', *Philosophical Magazine*, no. 14, 1907, pp. 749–60 (esp. p. 749). However, the Earth has been referred to as a greenhouse since the nineteenth century: see Augustin Mouchot, *La Chaleur solaire et ses applications industrielles*, Paris: Gauthier-Villars, 1869, pp. 2–3.

32 Fourier, 'Remarques générales', pp. 161–2.

alterations that extend over vast territories, *as a result of human works* or through the action of nature alone. These causes progressively impact on the average temperature of climates. The results cannot be other than tangible'.[33] Fourier certainly provided the theoretical formulation of an anthropogenic climate change, but this change was the one feared by his age: that generated by man in deforesting the Earth. In this sense, his works do not so much herald the theory of global warming as propose an initial mathematical reflection on the climatic effects of forests, desiccation, and cultivation.

The same may be said of John Tyndall. As shown by the historian Joshua Howe, the link he established between CO_2 and global temperature, which seems so fundamental to us today, was entirely marginal in his work. The topic only appears in two brief asides in the 'seminal' article of 1861,[34] never to recur for the duration of Tyndall's long career.[35] The heart of his work bore on thermodynamics and the relations between heat, motion, electricity and magnetism – in short, the major problems of the physics of his time. His principal innovation was experimental: the photospectrometer he invented served not so much to measure the greenhouse effect peculiar to different gases, as to precisely quantify the conversion factors between light, heat, and electricity.

Tyndall, Svante Arrhenius, Thomas Chamberlin, and other more obscure 'precursors' of the greenhouse effect also had another objective when they considered global temperatures: explaining the alteration of climates over geological time – both the great ice ages and the warm climates of the Tertiary Period.[36] Multiple hypotheses were invoked: intermittent weakening of solar radiation, the Earth's passage through 'cold' regions

33 Fourier, 'Extrait d'un Mémoire sur le Refroidissement séculaire du globe terrestre', *Annales de chimie et de physique*, vol. 13, 1820, pp. 418–38 (here, p. 435).

34 John Tyndall, 'On the Absorption and Radiation of Heat by Gases and Vapours, and on the Physical Connection of Radiation, Absorption, and Conduction', *Philosophical Transactions of the Royal Society of London*, vol. 151, 1861, pp. 1–36.

35 Joshua Howe, 'Getting Past the Greenhouse: John Tyndall and the Nineteenth-Century History of Climate Change', in B. Lightman and M. Reidy, eds, *The Age of Scientific Naturalism: Tyndall and His Contemporaries*, London: Pickering & Chatto, 2014, pp. 33–49.

36 Eunice Foote, 'Circumstances Affecting the Heat of the Sun's Rays', *American Journal of Science and Arts*, vol. 22, 1856, pp. 382–3; Svante Arrhenius, 'On the Influence of Carbonic Acid in the Air upon the Temperature of the Ground', *Philosophical Magazine*, series 5, no. 41, 1896, pp. 237–76; Thomas C. Chamberlin, 'A Group of Hypotheses Bearing on Climatic Changes', *Journal of Geology*, no. 5, 1897, pp. 653–83.

of interstellar space, modification of its axis of rotation following major geological upheavals, continental drift altering the terrestrial albedo, and so forth. Two theories in particular were discussed: an *astronomical* theory, attributing ice ages to alterations in the Earth's trajectory, and an *atmospheric* theory, pointing to natural alterations in its gaseous envelope and the effects of this on temperatures. Tyndall reckoned that his experiments strengthened the second hypothesis. Likewise Arrhenius, in an endlessly cited article, calculated the increase in the annual average temperature brought about by an increase in the concentration of CO_2 (for example, at 50° latitude it would be +3.65° C, or a 50 per cent rise in CO_2). But it was not in order to worry about man's impact on the climate. Arrhenius wanted rather to account for the warm climates of the Tertiary Period, when elephants and rhinoceroses gambolled even in the polar regions.[37]

This does not mean that the question of human action was altogether absent from these scientists' intellectual horizon. In *L'Évolution des mondes* (1910), Arrhenius thus registered in a footnote the rapid growth in global coal consumption (510 million tonnes in 1890, 1,209 million in 1907).[38] But it was cause for rejoicing: warming would afford 'the human race more equal temperatures and milder climatic conditions . . . which will enable the soil to yield significantly better harvests than today, for the good of a population that seems to be increasing faster than ever.'[39] Similarly, the meteorologist Nils Eckholm explained that coal consumption would make it possible 'to regulate the future climate of the Earth, and therefore to prevent the advent of a new ice age.'[40] In 1936, Guy Callendar, whose diagnosis came close to contemporary conclusions, could still rejoice at his discovery: the rise in CO_2 would increase harvests and, above all, 'the return of the deadly glaciers should be delayed indefinitely.'[41] The fearful anthropocene of the water cycle, the French Revolution and the empires was replaced by confidence, from the late nineteenth century to the 1960s,[42] in

37 Arrhenius, 'On the Influence of Carbonic Acid', pp. 268–9.

38 Svante Arrhenius, *L'Évolution des mondes*, Paris: Librairie Polytechnique, 1910, p. 58.

39 Ibid., p. 69.

40 Quoted in Fleming, *Historical Perspectives on Climate Change*, p. 111.

41 Guy Stewart Callendar, 'The Artificial Production of Carbon Dioxide and Its Influence on Temperature', *Journal of the Royal Meteorological Society*, vol. 64, 1938, pp. 223–40. On Callendar, see James R. Fleming, *The Callendar Effect: The Life and Work of Guy Stewart Callendar*, Boston: American Meteorological Society, 2007.

42 In 1966, Roger Revelle was still relatively optimistic: 'The increase in CO_2 should elicit more curiosity than fear.' Cf. history.aip.org/climate/Revelle.htm.

the stability of the CO_2 cycle – sometimes including the hope for an actual softening of the climate by means of carbon emissions.

The history of the precursors of climate change contains a doubly positivist view of contemporary environmental issues: it gives the impression that the climate alert flowed spontaneously from advances in science, and that climate change was 'discovered' by a handful of heroic great scientists. In short, global warming had to have its Galileo.

Yet the history of global change is in no way that of a 'discovery'. In the mid-1950s, American atomic scientists like Gilbert Plass, Edward Teller and Harrison Brown alerted both the public and the oil companies to the climatic consequences of increased CO_2 in the atmosphere. Exxon and Shell took these warnings seriously and immediately carried out some in-depth research, that was quickly buried.[43] What initially emerged as a curiosity and a remote threat became an increasingly incontestable scientific finding, thanks to the linkage between glacier climatology, advances in the study of the carbon cycle (isotopes played a major role), and the first computer simulations of atmospheric circulation.[44] This research was spurred by the United States which, engaged in the Cold War, had promoted the knowledge of the Earth's physical environment (continents, oceans, atmosphere) to the rank of strategic objective. The planet needed to be mapped, surveyed and modelled in order to be mastered as a space for the movements of ballistic missiles and submarines. The possibility of climate change was taken particularly seriously on account of its impact on the ice at the North Pole, the prospective battlefield of the Third World War.[45] The resources invested in the sciences of the terrestrial environment (and meteorology in particular) were multiplied tenfold.[46]

It was spotter aircraft, weather balloons, thermal buoys and satellites, plus the logistics enabling the organization, standardization and calibration of all these measures, which gradually made it possible to understand

43 Gilbert Plass, 'The Carbon Dioxide Theory of Climatic Change', *Tellus*, vol. 8, 1956, pp. 140–54. The *New York Times* reported this article (28 October 1956). Benjamin Franta, 'Early Oil Industry Knowledge of CO_2 and Global Warming', *Nature Climate Change*, no. 8, 2018, pp. 1024–5.

44 Spencer R. Weart, *The Discovery of Global Warming*.

45 Ron Doel, 'Quelle place pour les sciences de l'environnement physique dans l'histoire environnementale?', *Revue d'histoire moderne et contemporaine*, vol. 56, no. 4, December 2009, pp. 137–64.

46 The meteorological department of the US Air Force, for example, had 11,500 employees in the 1950s.

the evolution of the global climate. The historian Paul Edwards encapsulates this scientific endeavour with a striking image: proving climate change, he says, is like reconstructing a film on the basis of millions of photographs taken by different photographers and cameras. This enormous observational effort was soon assisted by supercomputers and by ever-more powerful, refined, and realistic global computer models – models around which whole communities of scientists were organized, each of them contributing their own stone to the edifice. In this sense, the diagnosis of global warming was not a 'discovery' but a finding, all the more robust for emerging out of a colossal collective effort of observation, analysis and modelling. It could only be arrived at thanks to an immense infrastructure of knowledge combining human beings and machines.

None of it would have been possible, moreover, had this expertise not overlapped with, first, the promotion of nuclear energy by the powerful Atomic Energy Commission and, second, with a growing preoccupation with the destructive effects of human activity on the planet in the 1960s and '70s.[47] Just like scientific knowledge, environmental reflexivity was embedded in the Cold War context. People were concerned about the effects of nuclear testing on the atmosphere and a possible 'nuclear winter' following an atomic exchange. Plans for 'meteorological warfare' and climate engineering were a source of fascination and terror.[48] All this fuelled in part the rise of the environmental movement and its fight against air pollution, supersonic aircraft, and damage to the ozone layer. The global atmosphere became a political issue that agitated scientists and activists, but also political leaders and public opinion.

Compounding this was the famine experienced by the countries of the Sahel between the late 1960s and the mid-1970s. Washington now feared that the scenario of the domino theory, the swing to communism that had loomed over Asia, might ultimately unfold in Africa instead. Experts and scientists were mobilized. One question concerned the causes of the famine.[49] According to some specialists on the atmosphere, the Sahel populations themselves were responsible for their misfortune, having

47 Jean-Baptiste Fressoz, 'La transition énergétique de l'utopie atomique au déni climatique', *Revue d'histoire moderne et contemporaine*, 2022, vol. 69, no. 2, pp. 114–46.

48 James R. Fleming, *Fixing the Sky: The Checkered History of Weather and Climate Control*, New York: Columbia University Press, 2010.

49 Fabien Locher, 'Third World Pastures: The Historical Roots of the Commons Paradigm (1965–1990)', *Quaderni Storici*, no. 1, April 2016, pp. 303–33.

wrecked the vegetation and soil with unsuitable agricultural and pastoral practices. Thus they had contributed to the drought, either by injecting a massive quantity of dust into the atmosphere or by acting on the region's albedo.[50] Like pollution or the ozone hole, the Sahel crisis helped make human climate action a key question even before global warming became established, from the late 1970s, as the great challenge facing humanity.

50 See, in very different styles, the interventions of Reid A. Bryson and Jule G. Charney. Reid A. Bryson and Thomas J. Murray, *Climates of Hunger: Mankind and the World's Changing Weather*, Madison: University of Wisconsin Press, 1977; Jule G. Charney, 'Dynamics of Deserts and Drought in the Sahel', *Quarterly Journal of the Royal Meteorological Society*, 101 (428), 1975, pp. 193–202. On Bryson, see Robert Luke Naylor, 'Reid Bryson: The Crisis Climatologist', *WIREs Climate Change*, vol. 13, no. 1, 2022.

Conclusion

If global warming came as a shock to most, and remains one, it is because since the beginning of the twentieth century, industrial civilization and science have instilled in us two convenient, but false, ideas. On the one hand, that human action cannot disrupt the climate; on the other, that wealthy societies no longer have very much to fear from its fluctuations. Our stupefaction when faced with the existential crisis of global warming largely stems from the reassuring illusions of a climate at once unshakeable and harmless. Such is the grim epilogue to the history retraced in this book.

At the end of the nineteenth century, Europe's climatic horizon brightened. The triumph of the train and the steam ship made people less vulnerable to the vagaries of the sky. The expansion of agricultural production and the globalization of supply broke the classic, recurrent linkage between meteorological events, poor harvests, panics, food shortages and famine. Richer countries were protected from the whims of a sun that always shone somewhere when it came to harvests. Climate changes – whether feared or desired – lost their political centrality as a result. At the same time, forests were no longer a political bone of contention hotly disputed by the assemblies of the past century. With coal and steel, trees no longer seemed to determine the fate of nations, armies, fleets, industry, and the public finances. The causes of droughts or floods were certainly considered and debated, but the impact of human activity on the climate was ever more seldom blamed for such disasters.

The debates that raged throughout the nineteenth century on the reality and causes of climate changes, on a scale of centuries or decades, had run out of steam. No side had won the battle: no decisive experiment, no irrefutable evidence, had demonstrated the stability of climates or their transformation, whether caused by man or not. Official climatological science often promoted descriptive approaches that repudiated the hypothesis of anthropogenic action or set it to one side. The threat had also been overplayed: instrumentalized in political fights to cast opprobrium on opponents, abused by administrators and foresters who had made it a tool of their power. All the while, knowledge of the Earth's history progressed: yes, the climate had changed, enormously even, but on the scale of geological time. Our planet had been torrid in the distant past and had experienced ice ages. Humanity now lived in the warmth of a period tailor-made for telling its epic story: the Holocene. Climate developments over thousands of years were discussed, as were the shorter cycles that punctuated human history. But the hand of the legislator, of the ruler, of the engineer no longer trembled at the thought of the climatic consequences of their actions.

The idea remained influential for longer in the empires, serving to indict the natives and their destructive ways of life. Colonization promised protection and reparation: reason, force, planning, and the white man's strict but paternal yoke were the guarantors of the existence of rain itself. The threat of anthropogenic climate change finally disappeared at the same time as the European empires of Africa and Asia. The new global imperium – that of US power – had its own obsession: the destruction of soils, already experienced on a large scale in the 1930s with the Dust Bowl. Many past empires had collapsed because they had exploited the land to excess: might the same happen to the United States? After 1945 these fears were transferred to the Third World, where a dearth of agricultural soil was feared. This fostered the rise of environmentalism and of developmental policies.

The evanescence of the vision of a fragile climate thus took nearly a century, from the train making its way through the European countryside to the missions of UN experts in the arid regions of the Third World. Thus ended four centuries of an initially optimistic, then anxious, Anthropocene. For a brief moment, man ceased to discern in the mirror of the sky and climate the reflection of a nature ready to rebel. This was an interlude: a loosening of the often oppressive ties which, since the dawn of the modern age, had bound European societies and their empires to the destinies of a fragile climate.

The loosening was twofold: not only were human societies no longer thought to act on the climate, but the latter was itself losing its powers of agency. In the late nineteenth century, new fields of knowledge emerged that saw life in terms of non-climatic causalities. Microbiology, for example, provided doctors with precise culprits and invalidated the direct links between environment and health. Likewise, with advances in genetics, heredity was severed from environment and its mechanisms withdrew to the heart of the cell and later into DNA. The climate was no longer the matrix of all life that had transfixed scientists for so long. The social sciences underwent similar changes. Sociology, then in its infancy, was constructed – especially in France – against naturalistic explanations. In economics, the 'neo-classical' current, later dominant, focused on the notion of utility and on mathematical methods, excluding natural determinants. In short, just as Western societies were becoming less vulnerable, new explanatory orders short-circuited the climate issue and reinforced the comfortable, fleeting sense of detachment sometimes called modernity.

Our amazement at global warming and our procrastination over the painful measures it requires stem in part from this dual heritage. During a brief interlude combining technical breakthroughs, scientific uncertainty, ice ages, globalization, microbes and social science, the climate was dislodged from our consciousness. We are still living in that epilogue, even though we no longer have the luxury of taking the climate for granted.

Afterword

This book arose from a find and a surprise: nearly 200 years ago, the French government launched a major inquiry into climate change and the human responsibility for this baleful process. Studying in detail the responses we dug up from the archives, it gradually emerged that the climate anxiety of which they were the manifest trace was actually not prophetic, nor even strange for the period.

The idea of the impact of human action on the climate was very widespread at that time among the French elite. It was also heir to a long history going back to the European colonization of the American continent. The Revolution and the political struggles of the nineteenth century had imparted to it a different, almost apocalyptic sense. While it was not a matter of CO_2 or the greenhouse effect, it addressed problems that were fundamental all the same: forest management, disruptions of the water cycle, the fate of mountains, the menace of flooding, the spectre of agricultural failure and political rebellion. The 1821 inquiry was but a tiny part of the vast ensemble of reflections, debates and scholarship on climate change that had permeated Europe, North America and the colonial empires, from India to the Sahel, for five centuries.

The subject is gigantic, and it has taken many years of investigation to get on top of it, accruing multiple debts along the way. The first is to the French public service and to our institution, the Centre national de la recherche scientifique (CNRS) which, despite the serious threats facing both today, made it possible for this book to exist. A long-haul undertaking like this one, with its patient effort of documentary collection and analysis, needs time. Our permanent status as civil servants and researchers, and the autonomy it affords us, have allowed us this time.

Our gratitude also goes to the people with whom we have had the opportunity to exchange views over the years and who have nurtured our thinking. To Dominique Pestre, first of all, who trained both of us in the history of the sciences. This book bears his unmistakeable imprint – the mark of a type of historiography that takes bodies of knowledge, their construction and materiality, seriously; that refuses retrospective judgement; and that seeks to resituate them in the political and social contexts of their time. Thanks also to Fredrik Albritton Jonsson, Monette Berthomier, Christophe Bonneuil, Samir Boumediene, Deborah Coen, Amy Dahan, Diana K. Davis, Michel and Josette Fressoz, Jérôme Gaillardet, Jean-François Gauvin, François Jarrige, Claude and Chantal Locher, Sara Miglietti, Séverine Nikel, Nicole Perrier, Caroline Pichon, Grégory Quenet, Marion Rampal and Martin Sarrazac, Simon Schaffer, Archie Shepp, Julien Vincent, Jean-Paul Zuniga, and our editor, Patrick Boucheron, for their advice, criticisms and encouragement. This book owes a great deal to all of them. And thanks to the Huntington Library for its research grant.

Our intellectual universe has changed considerably over the last ten years: trained as historians of the sciences, we have been protagonists in the emergence in France of the field of environmental history. The mounting ecological stakes in our societies, and the radical changes these entail for social-scientific investigation, have marked our lives as researchers and the pages of this book. Our thanks also go to our comrades in the Environmental History research group at the École des hautes études en sciences sociales (EHESS), with whom we have shared this experience of an emergent intellectual domain, still rich in themes to be explored.

We made the final corrections to this book during the lockdown of Spring 2020, in troubled times, marked by the return of certain figures familiar to the historian of past centuries: the epidemic as a reflection of our social and environmental anxieties; the forest as a bastion, a chance for salvation; and the fear of collapse, again and as ever, as a frightening horizon. Nature changes quickly, but our ways of thinking about that change draw on a long past made of discord and learning, part of whose history is told in this book.

Index

ill. denotes illustrations